Emerging Issues in Green Criminology

Critical Criminological Perspectives

The Palgrave *Critical Criminological Perspectives* book series aims to showcase the importance of critical criminological thinking when examining problems of crime, social harm and criminal and social justice. Critical perspectives have been instrumental in creating new research agendas and areas of criminological interest. By challenging state defined concepts of crime and rejecting positive analyses of criminality, critical criminological approaches continually push the boundaries and scope of criminology, creating new areas of focus and developing new ways of thinking about, and responding to, issues of social concern at local, national and global levels. Recent years have witnessed a flourishing of critical criminological narratives and this series seeks to capture the original and innovative ways that these discourses are engaging with contemporary issues of crime and justice.

Series editors:

Professor Reece Walters
Faculty of Law, Queensland University of Technology, Australia

Dr. Deborah Drake
Department of Social Policy and Criminology, The Open University, UK

Titles include:

Kerry Carrington, Matthew Ball, Erin O'Brien, Juan Tauri
CRIME, JUSTICE AND SOCIAL DEMOCRACY
International Perspectives

Claire Cohen
MALE RAPE IS A FEMINIST ISSUE
Feminism, Governmentality and Male Rape

Deborah Drake
PRISONS, PUNISHMENT AND THE PURSUIT OF SECURITY

Margaret Malloch, William Munro (*editors*)
CRIME, CRITIQUE AND UTOPIA

Maggi O'Neill and Lizzie Seal (*editors*)
TRANSGRESSIVE IMAGINATIONS
Crime, Deviance and Culture

Reece Walters, Diane Solomon Westerhuis, Tanya Wyatt (*editors*)
EMERGING ISSUES IN GREEN CRIMINOLOGY
Exploring Power, Justice and Harm

Critical Criminological Perspectives
Series Standing Order ISBN 9780-230-36045-7 hardback
(*outside North America only*)

You can receive future titles in this series as they are published by placing a standing order. Please contact your bookseller or, in case of difficulty, write to us at the address below with your name and address, the title of the series and the ISBN quoted above.

Customer Services Department, Macmillan Distribution Ltd, Houndmills, Basingstoke, Hampshire RG21 6XS, England

---

Emerging Issues in Green Criminology

Exploring Power, Justice and Harm

Edited by

Reece Walters
Assistant Dean, Research for the Faculty of Law at Queensland University of Technology, Australia

Diane Solomon Westerhuis
Independent Scholar

and

Tanya Wyatt
Senior Lecturer in Criminology, Northumbria University, UK

palgrave
macmillan

Softcover reprint of the hardcover 1st edition 2013 978-1-137-27398-7

First published 2013 by
PALGRAVE MACMILLAN

Palgrave Macmillan in the UK is an imprint of Macmillan Publishers Limited, registered in England, company number 785998, of Houndmills, Basingstoke, Hampshire RG21 6XS.

Palgrave Macmillan in the US is a division of St Martin's Press LLC, 175 Fifth Avenue, New York, NY 10010.

Palgrave Macmillan is the global academic imprint of the above companies and has companies and representatives throughout the world.

Palgrave® and Macmillan® are registered trademarks in the United States, the United Kingdom, Europe and other countries

ISBN 978-1-137-27397-0 ISBN 978-1-137-27399-4 (eBook)
DOI 10.1057/9781137273994

A catalogue record for this book is available from the British Library.

A catalog record for this book is available from the Library of Congress.

Contents

List of Illustrations

Figures

Tables

Boxes

Notes on Contributors

Lieselot Bisschop is a researcher at the research group Governing and Policing Security (GaPS) of the Department of Business and Public Administration at the University College Ghent, Belgium.

Avi Brisman is Assistant Professor in the School of Justice Studies at Eastern Kentucky University, United States.

Matthew Hall is Senior Lecturer in Law and Criminal Justice at the University of Sheffield, United Kingdom.

M. H. A. Kluin is a researcher at the Faculty of Technology, Policy and Management at Delft University of Technology, The Netherlands.

Olga Knight is a researcher at the University of Colorado, Denver, United States.

Peter Martin is a senior research associate in the Faculty of Law at Queensland University of Technology, Australia.

Hanneke Mol is a PhD candidate studying for a joint degree at the University of Kent, United Kingdom, and Utrecht University, The Netherlands.

Angus Nurse is a lecturer at Birmingham City University, United Kingdom.

Ragnhild Sollund is Professor in the Department of Criminology and Sociology of Law at the University of Oslo, Norway.

Nigel South is Professor and Pro-Vice-Chancellor (Academic Partnerships) in the Faculty of Law and Management at the University of Essex, United Kingdom.

Paul B. Stretesky is Professor in the School of Public Affairs at the University of Colorado, Denver, United States.

Gudrun Vande Walle is Assistant Professor at the University College Ghent (affiliated with Ghent University, Belgium) and Visiting Professor at Rühr University of Bochum, Germany.

Reece Walters is Professor of Criminology in the School of Justice and Assistant Dean Research, Faculty of Law at Queensland University of Technology, Brisbane, Australia.

Diane Solomon Westerhuis is an independent researcher and was previously a Lecturer of Criminology at Charles Sturt University, Bathhurst, New South Wales, Australia.

Rob White is Professor at the University of Tasmania, Australia.

Tanya Wyatt is Senior Lecturer in Criminology in the Department of Social Sciences at Northumbria University, Newcastle-Upon-Tyne, United Kingdom.

Introduction

Tanya Wyatt, Diane Solomon Westerhuis and Reece Walters

Professor Nigel South, one of the pioneers of green criminology, rightly observes that emerging environmental harms and injustices require 'a new academic way of looking at the world but also a new global politics'. This includes an intellectual discourse that moves 'beyond the narrow boundaries of traditional criminology and draws together political and practical action to shape public policy' (South, 2010: 242). Green criminology continues to evolve as a dynamic body of knowledge of resistance and innovation, one that challenges mainstream crime discourses and critically examines the policies and practices of contemporary governments and corporations. It is a collection of new and thought-provoking voices within the criminological lexicon, and its engagement with diverse narratives seeks to identify, theorise, and respond to environmental issues of both global and local concern. The expansion of green criminological perspectives serves to harness and mobilise academic, activist, and governmental interests to preserve, protect, and develop environmental issues.

This edited collection brings together scholars to explore green criminology through interdisciplinary lenses of power, justice, and harm. The chapters provide insightful case study analyses from North America, Europe, and Australia that seek to advance theoretical, policy, and practice discourses about environmental harm. This book also brings together transnational debates in environmental law, policy, and justice. In doing so, it examines international agreements and policies within diverse environmental discourses of sociology, criminology, and political economy. The contributors comprise a mixture of experienced and widely published scholars and those who have recently completed doctorates. This was quite a deliberate strategy to involve green criminological colleagues at different points of their careers with differing academic, policy, and activist backgrounds. All contributors have recently delivered papers at the 'York National Deviance Conference', the 'British Society of Criminology Conference', the 'European Society of Criminology Conference', the 'European Group', the 'Australian and New Zealand Society of Criminology Conference', the 'American Society of Criminology', and the inaugural 'Environmental Crime and Its Victims Conference' dedicated to green criminology in Delft, the Netherlands. The collection, therefore, represents a range of original, international, and cutting-edge works in the area and seeks to consolidate into one volume the most recent developments in this burgeoning field.

As indicated, the content encompasses material from a range of authors for a range of audiences, from those exploring the argument and debates of green criminology for the first time, to those seeking serious interrogations of pressing and demanding environmental problems. It draws on original and unpublished research and provides innovative insights into the detection, regulation and enforcement of environmental crimes. The book has a distinctive edge, notably its desire to integrate the concepts of power, justice, and harm into thinking about green criminological issues. To date, the growing body of green-centred criminological discourse has been events-based or focussed on case study–specific issues. This is, naturally, crucially important. As Nigel South stated back in 1998, it was essential for the foundations of a green criminology to be established through an examination of first, 'regulation, disasters and violations; second, legal and social censures; and third, social movement and environmental politics' (445).

It is evident that at the time Nigel South provided a platform that he and others have built on during the last 15 years. In our view, it is now time to begin thinking a little more conceptually and exploring what is theoretically distinct about green criminology. To this end, we invited contributors for this collection to focus on the ways that power, justice, and harm intersect with their chosen topic of inquiry. In a similar fashion to Hall and Winlow's (2012) endeavour to explore new horizons for critical criminological scholarship, it is our intention here to examine whether our chosen concepts provide a useful lens through which to craft deeper theoretical engagement.

The issues contained in this volume and others that explore green criminological issues often involve discussions of state and corporate power. We see that much criminological writing is about injustice and powerlessness; therefore, we begin by unravelling the relevance of power, injustice, and discrimination. This approach mirrors the work of Paul Gilroy's (1987) classic text *There Ain't No Black in the Union Jack,* which focussed on institutional bias and state-mobilised prejudice. The chapters throughout this book bind these topics with a conceptual twine made from debates about power, justice, and harm. First, power in its many facets is portrayed in many chapters. We witness 'coercive power', or 'power over', or what Bachrach and Baratz referred to as the mobilisation of bias (1970: 43). Here we observe individuals in positions of state and corporate authority systematically reinforcing predominant values through institutional practice. This is evident in Kluin's exploration in this volume of environmental regulators of the chemical industry and their power relationship with those who are regulated in the Netherlands. As demonstrated, those who benefit are placed in a preferred position to defend and promote their vested interests.

At another level, this book identifies various forms of influential power. Here networks of position are used to manipulate pressure and sway specific events for desired outcomes. The bribing of political officials, the falsification of documentation, flexing diplomatic or market muscle, and setting political agendas are all brought about by having the resources and 'strategic power' to influence outcomes (Walters, 2011). Influential power often operates in partnership with the subject or object of influence, unlike coercive power, which denies the disempowered access to processes, practices, and decisions. In addition, the book explores notions of 'soft power'. This is a form of 'ambassadorial power' based on charm, persuasion, and entrepreneurism. Here, those in elite or influential positions are able to present, construct, and promote their desired outcome through methods and images that appear to favour multiple parties. The images of corporate green responsibility and environmental regeneration projects from oil companies are persuasive ways to curry favour with sceptical and critical voices. The soft power model is what Tombs and Whyte (2010: 156) refer to as a form of 'covert power' in shaping nonconflict compliance: 'It works most effectively where it does not seem to be present at all'. Furthermore in the struggle to maintain or gain power, conflict arises, and this volume also addresses the role that environmental commodities have in the creation and continuance of such power-related conflicts.

Power is not a discrete category of exploration within green criminology. The dynamics related to power inevitably raise concerns around injustice. Definitions and conceptions of justice are essential to green criminology, as they provide the theoretical underpinnings and methods of green criminological research and debate. People's concept of justice will determine which acts or omissions they deem harmful or criminal, who (human, nonhuman animal, plant, or environment) they believe can be a victim, and how they approach their research design. Indeed the expanded notions of justice and harm, as are demonstrated in multiple chapters of this book, are what set green criminology apart from mainstream and conventional criminology. Justice is not confined to situations where the law has been properly adhered to or administered; justice is achieved when the rights of the people, the environment, and, depending on your justice framework, the rights of other species have been upheld. Additionally, concepts of justice influence the research methods used – for example, when researchers design projects that are socially and culturally inclusive (White, 2011). White (2011: 34) has categorised justice into three orientations that inform the green criminological discourse. They are:

> *Environmental justice* – in which environmental rights are seen as an extension of human or social rights so as to enhance the quality of human life, now and into the future.
>
> *Ecological justice* – in which it is acknowledged that human beings are merely one component of complex ecosystems that should be preserved for their own sake via the notion of the rights of the environment.
>
> *Species justice* – in which harm is constructed in relation to the placement of nonhuman animals within environments and their intrinsic right to not suffer abuse, whether this be one-on-one harm, institutionalised harm, or harm arising from human actions that affect climates and environments on a global scale.

On some level, the three orientations can be seen as the evolution of rights within green criminology, where at first rights for humans were extended to include the health of their environments. This was an important step in recognising the often discriminatory nature of pollution and other environmental degradation, but it still remained focused on humans as the only legitimate victims. Rights then expanded further to include the environment itself as having intrinsic value and then expanded again to include other species having the right to life and freedom from suffering. These non-anthropocentric concepts of justice are fundamental to the uniqueness of green criminology and to its ability to enhance the criminological gaze. These orientations guide green criminologists in problematising once-invisible injustices that were occurring with the degradation and destruction of the environment.

Whilst they can be, it is not essential that these categories be viewed as discrete, where the concepts of harm and rights are only within one or another of the orientations. It is possible to regard a healthy environment as a human right while also believing that nonhuman animals should not be routinely abused. It is also possible to see people as part of the larger ecosystem while believing that the ecosystem and the species within it have rights.

The next step for conceptualising justice within green criminology is to develop frameworks in which competing interests are addressed and in some way balanced. There is a long way to go in theorising justice in regard to environmental, nonhuman animal, and human well-being, as will be discussed here in various ways. Additionally, more horizon scanning needs to take place that will predict future injustices, most likely stemming from catastrophes and conflicts (South, 2010), and will take into account intergenerational justice, where leaving a healthy, diverse environment for those not yet born is taken into consideration when examining the harmfulness of actions or omissions (South, 2009). Whilst there is movement within green criminology to tackle the global aspects of green crimes, further development of conceptions of justice that capture the complexity of transnational justice that stems from the interconnectedness of the local to the global levels must also progress (Walters, 2007), and within this development there must be further continuance of the inclusive nature of green criminological research.

However, green criminology and this book not only theorise about definitions and conceptions of justice as described above; green criminological research also unpicks the functioning of the environmental justice system itself. In instances where environmental laws have been violated, it is important to analyse whether justice has been achieved. Several chapters in this collection explore the role of actors such as law enforcement, the courts, and international nongovernmental organisations in making sure that the victims receive justice when an environmental crime has occurred. The types of harm and extent of harm that victims endure are also defined and conceptualised within the chapters. As the chapters will demonstrate, the concept of harm and the discourses defining and assessing harm in green criminology and environmental law are diverse, complex, and confusing. While environmental laws and regulations are designed to 'protect' the environment, in practice they work to assign value to harms already caused, in order to measure the consequences of some action, and thus measure and apply sanctions. To sanction harm, the law must assess penalties for inflicting harm, and to do this, the law must measure those harms to assess the level of sanction to apply. The necessity to legally assess harm and regulate sanctions has become necessary with the global increase in environmental law which, since the 1970s, has regulated and legislated outcomes of behaviour and actions that impact upon 'the environment' – another concept that is rarely defined and infrequently considered, other than to generalise about ecosystems. Such legal requirements have produced a theoretically robust narrative about ecological harm, environmental harm, and species harm, located differentially in diverse national legislation and in academic research, such as that produced by Halsey (2004) or White (for example, 2002 or 2008).

This volume not only contains explorations of how legislation and the criminal justice system deal with harm; it also delves into definitions and concepts of environmental harm that are outside the gaze of the legal system. Yet the notion of the environment itself is theorised much less often, and we find that the law in particular struggles to include any real definitions of the environment, or indeed of any prevention of harm to the environment, instead focussing on the impact of human actions, which cause harm to the health or business of humans from a very anthropocentric and economic perspective. In legislation and policy, a discourse of sustainability has reigned supreme. This discourse is all about allowing development, economic sustainability, and the continuance of human activities that may harm the environment and nonhuman species. The sustainability discourse has subsumed and overtaken any understanding of protecting the environment for moral or ethical reasons; instead it has been focussing on economic and developmental concerns. Examples of this economic and anthropocentric perspective abound, in United Nations declarations, in the laws of the United States and European laws. Diane Solomon Westerhuis (see Chapter 10) explores such definitions of harm in legislation and in the courts, citing examples of how harm is defined and measured in countries with high rankings of environmental impact.

The legal definitions we do have of harm and the environment are generally derived from United Nations conventions or treaties, and then transposed to diverse national and local legislation and regulations, which are intended to support our international agreements. In all these narratives, the extent to which environmental harm is considered is constrained to those harms that affect the resources and processes in natural ecosystems, but only those from which we are able to extract or gain benefit. Other impacts are rarely if ever considered or sanctioned. This anthropocentric focus is upon 'ecosystem services', a concept people have created that describes benefits to humans that can be obtained from the environment. These are then linked to risk, harm, and the future of food availability, rather than to any concerns about animals other than humans. As demonstrated in numerous chapters, when these other animals are considered, it is in the context of advantages to humans, or future generations of humans, or to the resources or services that they offer, albeit to their detriment or harm, in many cases. And so we come full circle: the concept of harm is rationalised, commodified, and anthropocentric. The epitome of such discourse appears in the United Nations 2005 *Millennium Ecosystem Assessment (MEA)*, which categorises ecosystem services and what they do to provide humans with benefits. The discourse is economic and rationalist and the categories are anthropocentric or human centred, forcing the environment to have both a structure and an agency dependent upon its ability to provision and support humans; thus, we have ecosystem services. This volume challenges such anthropocentric notions of harm as well as advancing concepts of justice, while uncovering power dynamics that lead to environmental degradation and harm.

So we see that the works here bring the concepts of power, justice, and harm to the forefront, using many different voices. In the following chapters, we find these concepts reappearing in diverse forms, in narratives and critiques that identify, theorise, and respond to a range of global and local environmental issues and that emphasise theoretically distinct approaches in green criminological discourse. Part I, 'Concepts, Perspectives, and Dimensions', begins with Rob White's contribution, 'The Conceptual Contours of Green Criminology'. This chapter outlines the origins and substantive focus of green criminology, pointing out that many researchers were working within the bailiwick of environmental crime without necessarily identifying their work as green criminology. As White reminds us, green criminology today includes studies of activities that are harmful but that may in fact be legal, and includes harms to humans, animals, and plants. White also identifies distinctive perspectives, including radical green criminology and eco-global criminology, through to conservation criminology and speciesist criminology. The chapter notes the commonalities and differences across the wide range of views expressed under the green criminology umbrella, before concluding that investigation of environmental harm incorporates many who travel the same road but who do so in their own particular ways.

Chapter 2, by Lieselot Bisschop and Gudrun Vande Walle, 'Environmental Victimisation and Conflict Resolution: A Case Study of e-Waste,' furthers this discussion. This study of victimisation that is suffered because of environmental crimes is relatively new, in that it takes up the idea of conflict resolution as a way forward for communities and individuals who have suffered harm and for the actors responsible. This chapter links environmental victimisation to conflict resolution, whether through retributive reactions or restorative justice; the latter is a notion that is gaining popularity and is also explored briefly by Diane Solomon Westerhuis in Chapter 10. Bisschop and Vande Walle use a case study of illegal transports and dumping of e-waste to further explore the development of environmental victimology.

This leads to Part II, 'Rights and Wrongs', beginning with Chapter 3 by Avi Brisman and Nigel South, 'Resource Wealth, Power, Crime, and Conflict'. Brisman and South describe linkages between environmental wealth and illegal trade in minerals, timber, and wildlife, and the internal conflicts, corruption, and flow of profits to external locations that can leave resource-rich countries with a poor return. Chapter 3 reviews ways in which natural resources can be a source of conflict and power, can fuel or fund existing conflicts, and can be a casualty of conflict. In these processes, the results include environmental degradation and the erosion or denial of social justice and human rights. The authors outline various approaches to amelioration, prosecution, and protection of populations and species, but agree that there is no single solution. Brisman and South conclude that in current global economic conditions there will be limits to the impact of democratic development in resource-producer nations unless Western production practices and exploitative patterns of consumption are addressed.

In Chapter 4, 'Animal Trafficking and Trade: Abuse and Species Injustice', Ragnhild Sollund uses a case study from Norway to highlight the unacceptability of harms caused by the growing trade in different animal species, an issue also raised by Tanya Wyatt in Chapter 6. For Sollund, the issues raised in Norway include threats to many groups of animals, such as parrots and reptiles, but the author notes these are threats that are common globally. The problems of illegal smuggling of animals and control of animal trafficking are, Sollund reminds us, poorly prioritised, punishment is lenient, and convictions are few. This chapter demonstrates problems of criminalisation versus legalisation, and contrasts trade and the keeping of different threatened species within both licit and illicit markets.

In Chapter 5, 'Crime and the Commodification of Carbon', Reece Walters and Peter Martin make connections between recent shifts in environmental practices and crime. They examine greenhouse gas emissions in the context of the law. They also follow predictions of increasing global temperatures and the rapid increase of worldwide carbon emissions, which continue to compromise environmental sustainability whilst contributing to premature death. Walters and Martin bring together recent global expansions of environmental law and regulatory arrangements with the introduction of new practices (stimulated by the Kyoto Protocol) of carbon taxes and sinks, and renewable or 'cleaner' energies, which produce commodities for global trade. The authors illustrate the power dimensions in the political economy of carbon and new opportunities for criminal enterprises, including fraud and the perpetration of environmental harms, a recurring theme in all chapters of this publication.

Tanya Wyatt studies 'The Local Context of Transnational Wildlife Trafficking: The Heathrow Animal Reception Centre' in Chapter 6. She suggests that wildlife trafficking most likely conjures up images of a faraway seedy exotic street market full of cages of wildlife bound for stew pots or a collector on the other side of the globe. The author takes us behind the scenes to compare such a scenario with just one reality – an animal reception centre at London's Heathrow International Airport. Wyatt describes the illegally trafficked wildlife that arrives regularly, including animals and plants that are protected under the Convention on International Trade in Endangered Species of Wild Flora and Fauna (CITES), who are seized and then housed in the Heathrow Animal Reception Centre (HARC). She describes in sometimes harrowing detail the nature and extent of seizures by the CITES enforcement authority, providing evidence of transnational green crimes and the harm caused on the local level.

Part III is 'Policing, Regulation, and Enforcement'. In Chapter 7, 'Perspectives on Criminality in Wildlife' Angus Nurse critiques criminal justice responses to wildlife law in Britain. Nurse explains that in the United Kingdom and the United States, nongovernmental organisations (NGOs) frequently argue for a strengthened wildlife enforcement regime with tougher sentences, but Nurse contests their perceptions of inadequate wildlife laws. Instead, Nurse argues that a law-enforcement approach fails to take into account the diverse nature of wildlife offending and criminality, because it is dominated by the perception of all wildlife offenders as rational-thinking, profit-driven individuals. The author offers a new typology of offenders, arguing that legislative changes and a more punitive regime are inadequate solutions to achieve justice in terms of wildlife crime problems, unless the existence of different types of offender and of criminal behaviour are recognised and addressed in policy and in enforcement practice.

M. H. A. Kluin also argues for a different approach in Chapter 8, 'Environmental Regulation in Chemical Corporations: Preliminary Results of a Case Study'. She responds to the call for more study of the role of street-level bureaucrats in the implementation of environmental regulations. Kluin's focus is on enforcement of regulation by field-level inspectors from an environmental protection agency, Seveso, in the Netherlands, describing violations of environmental regulations by chemical corporations. Kluin combines participant observations of Seveso inspections at four chemical corporations with their registered offences and enforcement activities, and she concludes that document analysis only shows the end result of these proactive and reactive enforcement activities, whereas participant observation gives a more complete view of what is really going in the field. The research provides insights into how regulators approach their interactions with regulatees, and how they ensure environmental compliance and ultimately justice.

Chapter 9, 'The Uneven Geography of Environmental Enforcement INGOs', by Paul B. Stretesky and Olga Knight, continues this study of enforcement. The authors claim that green criminologists have downplayed the role of nonstate actors in addressing environmental crime. Their research builds upon the 'treadmill of production theory' to examine the relationship between international nongovernmental organisations (INGOs) and income across countries. Specifically, Stretesky and Knight study the distribution of formalised INGOs that engage in environmental enforcement and related activities, including environmental crime-monitoring efforts and advocacy for stricter enforcement of environmental laws. The research includes data on INGOs, drawing upon the *Encyclopaedia of International Organizations*, the *Yearbook of International Organizations*, and the *World Directory of Environmental Organizations*. Stretesky and Knight find significant implications for ecological disorganisation and environmental enforcement within the context of environmental justice and the development of a global civil society.

Environmental justice is also of concern to Diane Solomon Westerhuis, who in Chapter 10, 'A Harm Analysis of Environmental Crime', undertakes to explore how environmental crimes are prosecuted and sanctioned. She describes prosecutions in the New South Wales Land and Environment Court (NSWLEC) in Australia. This study analyses the range of crimes prosecuted and the responses to environmental harms in this court since its inception in 1980. While some hope appears in a few court discussions about restorative justice, Solomon

Westerhuis draws the conclusion that sanctions are most frequently about punishment and deterrence, and that remediation for the harms inflicted was evident in little more than half the cases examined. She makes recommendations for change as a consequence of this research, in the hope of achieving some sort of environmental justice for victims and remedying harms inflicted. The analysis of the sanctions for these crimes demonstrates how sanctions have changed over time with our shifting perspectives of green crime.

In Chapter 11, Matthew Hall looks at 'Victims of Environmental Harms and Their Role in National and International Justice'. He focuses on the often-overlooked human victims of environmental degradation, examining the range of physical, social, and economic harms that can be attributed to environmentally destructive practices and then assessing the position of such victims within the national and international legal order, particularly with regard to criminal justice. Hall discusses how research and theory from the wider field of victimology might be applied and highlights the particular challenges posed by this form of victimisation within and beyond criminal justice. Hall calls for an interdisciplinary approach encompassing criminology, victimology, and international law, as well as other disciplines from the social and physical sciences, and argues the case for more firsthand empirical work in which environmental victims are questioned about their experiences and support needs. The closing chapter, Chapter 12 by Hanneke Mol, is '"A Gift from the Tropics to the World": Power, Harm, and Palm Oil'. Drawing on the case of oil palm cultivation in Colombia's South Pacific region, Mol examines the social and environmental harms associated with the growing global palm oil industry. Framed in a discussion of power vis-à-vis harm, the principal argument that Mol puts forward is that colonial forms of control, appropriation, and territorial ordering are intimately connected with the power relation between the human and the nonhuman. Such an approach accentuates the crucial role that green criminological perspectives assume in extending and deepening the central focus on power relations within critical criminology; it does this by turning the lens to the nonhuman, the intricate connection between the human and nonhuman realm, and by laying bare mechanisms of ordering *human* life through perspectives on and interventions in the natural world.

This book brings together a range of cutting-edge debates in green criminology in 'pursuit of social justice and human rights' (South and Brisman, 2013: 99). It weaves together notions of power, harm, and justice to examine emerging issues of environmental concern. Not so very long ago, such a book would not have been considered 'criminological'. 'But is it criminology?' queries Rob White, arguing that green criminology is of course an essential part of the criminological landscape. For he rightly identifies that issues of harm and – we would add – also of power and justice, have always been the domain of a critical criminology (White, 2013: 87). Interestingly, the questions 'What does this have to do with criminology?' or 'Why is this criminology?' are questions asked by all the contributors at varying times during the past decade after presenting at conferences; these questions seem to have subsided more recently. The need to justify the contents of this book within the criminological lexicon at conferences or within journals and books has now passed. The flourishing scholarship in journals, books, and conference presentations has served to embed discourses in green criminology in the broader crime and criminal justice landscape. Our intention here is to further the debate through this collection of excellent chapters. In doing so, we endorse the following words of David Garland (2013), with specific relevance to future green criminological endeavour:

> A criminology that aims to understand the facts of crime and justice in relation to the structures of political and economic power has to move from the particular to the general in ways that respect intermediating processes and mechanisms involved. If it aims to paint the big picture it has to combine, broad, impressionistic brushstrokes with pointillist factual detail and faithful attention to proportion, perspective and composition. And, as always in the portrayal of social life, it must seek to capture the play of irony and contradiction, and avoid the tendency to gloss these over in pursuit of a too-neat simplicity or order. (xi)

References

Bachrach, P. and Baratz, M. (1970) *Power and Poverty: Theory and Practice*. New York: Oxford University Press.

Garland, D. (2013) 'Forward', in K. Carrington, M. Ball, E. O'Brien and J. Tauri (eds) *Crime, Justice and Social Democracy – International Perspectives*. London: Palgrave.

Gilroy, P. (1987) *There Ain't No Black in the Union Jack: The Cultural Politics of Race and Nation*. London: Routledge.

Hall, S. and Winlow, S. (eds) (2012) *New Directions in Criminological Theory*. London: Routledge.

South, N. (1998) 'Corporate and state crime against the environment: foundations for a green perspective in European criminology', in V. Ruggiero,

N. South and I. Taylor (eds) *The New European Criminology, Crime and Social Order in Europe*. London: Routledge, pp. 443–461.

South, N. (2009) 'Ecocide, conflict and climate change: challenges for criminology and the research agenda in the 21st Century', in K. Kangaspunta and I. H. Marshall (eds) *Eco-crime and Justice: Essays on Environmental Crime*. Turin: UNICRI.

South, N. (2010) 'The ecocidal tendencies of late modernity: transnational crime, social exclusion, victims and rights', in R. White (ed.) *Global Environmental Harm: Criminological Perspectives*. Cullompton: Willan Publishing, pp. 228–247.

South, N. and Brisman, N. (2013) 'Critical green criminology, environmental rights and crimes of exploitation', in S. Winlow and S. Atkinson (eds) *New Directions in Crime and Deviance*. London: Routledge, pp. 99–110.

Tombs, S. and Whyte, D. (2010) 'Crime, harm and corporate power', in J. Muncie, D. Talbot and R. Walters (eds) *Crime: Local and Global*. Cullompton: Willan Publishing.

Walters, R. (2007) 'Crime, regulation and waste in the United Kingdom', in P. Beirne and N. South (eds) *Issues in Green Criminology: Confronting Harms against Environments, Humanity and Other Animals*. Cullompton: Willan Publishing.

Walters, R. (2011) *Eco Crime and Genetically Modified Food*. London: Routledge.

White, R. (2011) *Transnational Environmental Crime: Toward an Eco-global Criminology*. London: Routledge.

Part I

Concepts, Perspectives, and Dimensions

1 The Conceptual Contours of Green Criminology

Rob White

Introduction

Green criminology means different things to different people. Ostensibly 'green criminology' emerged in the early 1990s to describe a critical and sustained approach to the study of environmental crime. However, the tasks described by the term had in fact been undertaken well before the 1990s. The critical study of environmental crime thus predates green criminology as such.

Depending upon how 'critical' and 'sustained' are interpreted, green criminology has either a broad or a narrow purview. That is, those who view themselves as doing green criminology define it in ways that best suit their own conception of what it is they are doing. As seen below, there are periodic attempts to refine and redefine the green criminology project in both political and/or conceptual terms.

Nonetheless, the application of the term 'green criminology' is increasingly used to denote generic interest in the study of environmental crime, rather than a specific viewpoint in relation to these kinds of social and ecological harm. This is evident, for example, in the recent proliferation of 'green criminology' sessions at conferences of the American Society of Criminology, the British Society of Criminology, and the European Society of Criminology. Specialist conferences such as the 'Environmental Crime and Its Victims' conference held in Delft, the Netherlands, in September 2012 also incorporate a wide variety of viewpoints and substantive concerns, and are typically referred to as green criminology conferences.

This chapter provides an outline of the distinctive features of green criminology, its main concepts and foci of analysis, and the continuing debates that mark its further and continuing development as a *bona fide* perspective within criminology.

The origins and substantive focus of green criminology

There have been a number of 'fellow travellers' to green criminology, that is, scholars working on issues pertaining to environmental crime but not necessarily employing a green criminology perspective, which continues to this day. For instance, there is a body of work on environment-related issues that is informed by the literature on corporate and organised crime. Environmental harm and crime is linked to the activities of corporations and also to organised criminal syndicates, and analysis proceeds through the lens of sociology of deviance and political economy. Key concepts and concerns of these studies have included the control and manipulation of waste disposal processes, and the production and distribution of toxic chemicals (Pearce and Tombs, 1998; Ruggiero, 1996; Massari and Monzini, 2004).

Others have looked specifically at environmental crime, but generally within conventional frameworks. Here the focus has been on traditional illegal activities associated with the environment (such as, for example, illegal fishing), analysed within traditional criminological theoretical and practice approaches. The key concepts and concerns of this work have been based upon legal concepts of environmental crime, existing legislative and regulatory measures around environmental crime, and the nature of official environmental law enforcement (Situ and Emmons, 2000; Fyfe and Reeves, 2009; Shelley and Crow, 2009).

This criminological work overlaps to some extent with traditional legal studies in this area. The latter involve a conventional legal approach to the study of environmental crime as a violation of criminal law and civil statutes – basically legal studies with environmental crime as the object of analysis. Key concepts and concerns in this area of substantive law include notions such as vicarious liability, public interest, and attempts to place a legal/monetary value on environmental harm (Brickey, 2008; Bell and McGillivray, 2008; Mehta, 2009).

A fourth grouping has applied mainly sociological analysis to questions that pertain specifically to environmental justice. The main thrust of this work has been to explore the empirical links between toxic environments and certain categories of people (inevitably the poor, the dispossessed, and people of colour), and to actively struggle against the discrimination and racism that underpins such ecological injustice. The key concepts and concerns of this work have emphasised issues relating to the distribution of environmental advantage and disadvantage, rather than crime *per se* (Bullard, 2005; Pellow, 2007).

The early pioneers of a distinctive green criminology sought to provide a particular and self-conscious branding of the kind of work they engaged in. This period, in the 1990s, was characterised by writing about the need for criminology to take environmental crimes seriously, and to do so in ways that would force criminology to rethink how it does what it does and how to conceptualise the issues. Key concepts and concerns included the notion of green criminology itself as a concept; the idea that green criminology is a perspective, not a theory; and the social and ecological importance of studying environmental crime and harm (Lynch, 1990; South, 1998; Clifford, 1998).

Today, as a broad generic term, green criminology refers to the study by criminologists of environmental harms (which may incorporate wider conceptions of crime than that provided in strictly legal definitions), environmental laws (including enforcement, prosecution, and sentencing practices), and environmental regulation (systems of civil and criminal law that are designed to protect and preserve specified environments and species and to manage the negative consequences of particular industrial processes) (White, 2008, 2011).

The key focus of green criminology is environmental crime. This is conceptualised in several different ways within the broad framework of green criminology. For some writers, environmental crime is defined narrowly within strict legal definitions: It is what the law says it is. For others, environmental harm is itself deemed to be a social and ecological crime, regardless of legal status. If harm is done to humans or environments or animals, then it is argued that this ought to be considered a 'crime' from the point of view of the critical green criminologist.

Specific types of harm as described in law include things such as illegal transport and dumping of toxic waste, the transportation of hazardous materials such as ozone-depleting substances, the illegal traffic in real or purported radioactive or nuclear substances, the proliferation of e-waste generated by the disposal of tens of thousands of computers and other equipment, the safe disposal of old ships and airplanes, the illegal trade in flora and fauna, and illegal fishing and logging.

However, within green criminology there is also a more expansive definition of environmental crime or harm that includes (White, 2011):

- transgressions that are *harmful to humans, environments, and nonhuman animals*, regardless of legality per se; and

- environmental-related harms that are facilitated by *the state,* as well as *corporations and other powerful actors,* insofar as these institutions have the capacity to shape official definitions of environmental crime in ways that allow, condone, or excuse environmentally harmful practices.

What constitutes environmental crime is, therefore, contentious and ambiguous. Much depends upon who is defining the harm and what criteria are used in assessing the nature of the activities so described – for example, legal versus ecological, criminal justice versus social justice (see Situ and Emmons, 2000; Beirne and South, 2007; White, 2008).

Most green criminology is informed by at least one of the three approaches that collectively make up an eco-justice perspective (White, 2008). From an eco-justice perspective, environmental harm is best seen in terms of justice, which in turn is based upon notions of human, ecological, and species rights and broad egalitarian principles. Environmental victimisation is considered from the point of view of transgressions against humans, specific biospheres or environments, and nonhuman animals (and, increasingly, plants). This is conceptualised in terms of three broad areas of analytical interest:

- *environmental justice,* where the main focus is on differences within the human population; social justice demands access to healthy and safe environments for all, and for future generations
- *ecological justice,* where the main focus is on 'the environment' as such; to conserve and protect ecological well-being, for example forests, is seen as intrinsically worthwhile
- *species justice,* where the main focus is on ensuring the well-being of both species as a whole, such as whales or polar bears, and of individual animals, which should be shielded from abuse, degradation and torture.

Language intrinsically shapes how 'harm' and 'value' are constructed in regard to specific groups of humans, specific biospheres, and specific nonhuman species. For example, from a conservation criminology perspective (see for example, Gibbs et al., 2010a; Herbig and Joubert, 2006), the language used in referring to animals tends to be anthropocentric and instrumental. Thus, animals are categorised in terms of 'wildlife' and 'fisheries'. Environmental laws and laws specifically about animals likewise tend to define animals in ways that describe their existence and 'value' through reference to human conceptions and human uses (Sankoff and White, 2009; White, 2011). By contrast, those criminologists who write primarily about animal rights and animal welfare issues describe such anthropocentric language as a form of 'speciesism' (see Beirne, 2007; Sollund, 2008). From this perspective, it is the suffering of nonhuman animals – whether construed as wild, domestic or commercial – that is of central concern, not whether the suffering stems from illegal criminal acts or not, since much animal suffering is linked to legal activities such as abattoirs and factory farms that rely upon animals as food sources. Accordingly, the language employed is informed by animal-centred rather than human-centred considerations.

Major factors that influence study within green criminology, therefore, are the specific interests that count the most when conceptualising the nature and seriousness of the harm. For example, when criminalisation does occur, it often reflects human-centred (or anthropocentric) notions of what is best (for example, protection of legal fisheries or legal timber coupes) in ways that treat 'nature' and 'wildlife' simply and mainly as resources for human exploitation. The intrinsic value of specific ecological areas and particular species tends to be downplayed or ignored. Nonetheless, recent years have seen greater legislative and judicial attention also being given to the rights of the environment *per se*, and to the rights of certain species of nonhuman animals.

Drawing upon a wide range of ideas and empirical materials, green criminology has ventured across many different areas of concern. For example, it has documented the existence of lawbreaking and harmful activity with respect to air pollution, disposal of toxic waste, and misuse of environmental resources (Pellow, 2007; Gibbs, McGarrell and Axelrod, 2010b; Walters, 2011). Other work has examined the distribution of environmental 'risk', particularly as these affect poor and minority populations (Bullard, 2005), and has considered the specific place of animals in relation to issues of rights and human–nonhuman relationships on a shared planet (Benton, 1998; Beirne, 2009, 2011). Environmental victimisation is similarly a growing area for concerted analytical and practical attention (Williams, 1996; Hall, 2013).

Conceptual foundations and theoretical perspectives

Green criminology is premised on the idea that the justice system needs to take environmental harm seriously. For some exponents, this also means a need for conceptualisations of harm that go beyond conventional understandings of crime (Beirne and South, 2007). In the first place, green criminologists agree that destructive and damaging human activities that harm environments warrant greater attention than has hitherto been the case within criminology. There are, of course, already a plethora of laws and conventions that deal with environmental crime and environmental harms. Yet, until recently, very little criminological attention has been given to analysis of how these are actually working. Nor have criminological insights in areas such as crime prevention been rigorously applied to crimes involving animals and nature.

In the second instance, if ecological (and social and economic) welfare is to be maximised, then there is a need to expand notions of what actually constitutes environmental crime. Harm, as conceived by critical green criminologists, for example, demands more encompassing definitions than that offered by mainstream law and traditional criminology. This is because some of the most ecologically destructive activities, such as clear-felling of old-growth forests, is quite legal, while more benign practices, such as growing of hemp (an extremely strong fibre), is criminalised.

Green criminology provides an umbrella under which to theorise and critique both *illegal* environmental harms (that is, environmental harms currently defined as unlawful and therefore punishable) and *legal* environmental harms (that is, environmental harms currently condoned as lawful but which are nevertheless socially and ecologically harmful). How harm is conceptualised is thus partly shaped by how the legal–illegal divide is construed within specific research and analysis.

There is no green criminology *theory* as such. Rather, as observed by South (1998), there is what can loosely be described as a green 'perspective'. Elements of this perspective generally include things such as a concern with specifically environmental issues, social justice, ecological consciousness, the destructive nature of global capitalism, the role of the nation-state (and regional and global regulatory bodies), and inequality and discrimination as these relate to class, gender, race, and nonhuman animals. Corporate definitions of a green agenda are sometimes explicitly rejected (Lynch and Stretesky, 2003), insofar as corporations are generally seen to be integral to the problems of environmental harm. The green criminology perspective, therefore, tends to begin with a strong sensitivity toward crimes of the powerful and to be infused with issues pertaining to power, justice, inequality, and democracy.

Within the spectrum of ideas and activities associated with green criminology, there are several different kinds of analytical framework. Some of these pertain to eco-philosophy – that is, to ways in which the relationship between humans and nature can be conceptualised.

Academic work in this area includes consideration of gendered views of the natural and social worlds (Lane, 1998; Plumwood, 2005) and exploration of anthropocentric, biocentric, and ecocentric perspectives (Halsey and White, 1998) through to elaboration of postmodern versions of a constitutive green criminology (Halsey, 2004). Less abstractly, however, most environmental criminology can be distinguished on the basis of who or what precisely it is that is being victimised. This is represented in the three approaches that together constitute an eco-justice perspective, with their varying focus on humans, eco-systems, and animals.

Green criminology has emerged in the last 20 years as a distinctive area of research, scholarship, and intervention. It is distinctive in the sense that it has directed much greater attention to environmental crime and harm than mainstream criminology has done, and it has heightened awareness of emergent issues such as the problems arising from disposal of electronic waste (e-waste) and the social and ecological injustices linked to the corporate colonisation of nature, including biopiracy and imposition of genetically modified organism (GMO) crops in developing countries.

Perspectives within green criminology

As green criminology has grown as a specific area of concentrated scholarship and research so, too, it has developed distinct sub-areas or perspectives that express quite different conceptualisations of the problem and how best to respond to it. These are briefly summarised below (see Box 1.1). It's important to note that any one writer may be identified with more than one of the analytical streams listed here; the categorisations are not mutually exclusive.

The interface of criminology with environmental issues as a discrete field of study, in a manner that involves increased and concerted professional attention and hands-on intervention, has been forcefully advocated by Lynch and Stretesky (2003: 231).

Box 1.1 Perspectives within green criminology

Radical green criminology

Generic term to describe a *broad radical orientation* toward issues pertaining to *environmental harm* and crimes against nature.

Key concepts and concerns: ecological, environmental, species justice; anti-capitalist, anti-anthropocentric; environmentalist, animal rights

Exemplar: Lynch and Stretesky (2003), who provide a trenchant critique of corporations and who argue that green criminology ought to be defined precisely by its radical critique of the status quo.

Eco-global criminology

Main concern is with the *transnational nature of environmental harm* and the ways in which transgressions against humans, ecosystems, and animals manifest at a global level.

Key concepts and concerns: climate change; transnational environmental crime; ecological justice

Exemplar: White (2011), who argues that ecological criteria should underpin analysis, and that such analysis should be highly cognisant of the importance of scale, which incorporates the intersections of the local, national, regional, and international.

Conservation criminology

Designates a specific concern with *natural resource conservation* and management that draws upon criminological concerns, and with environmental law enforcement and *environmental crime as legally defined.*

Key concepts and concerns: conservation; natural resource management; risk assessment and analysis

Exemplar: Gibbs et al. (2010a), who argue for an integration of criminology with natural resource disciplines and the risk and decision sciences, so that study of environmental crimes and risks better incorporate interdisciplinary scholarship.

Environmental criminology

Conventional criminological approach to dealing with environmental crime as legally defined, drawing mainly upon place-based criminology (also known as 'environmental criminology') that concentrates on *situational crime prevention.*

Key concepts and concerns: situational crime prevention; market reduction approach; illegal wildlife trade

Exemplar: Wellsmith (2010), who argues that place-based criminology and situational crime prevention have much to offer in reducing environmental harm, especially in areas such as 'wildlife crime' and endangered species conservation.

Constructivist green criminology

Approaches the study of environmental harm and crime from the point of view of constitutive or constructivist criminology, which emphasises how categories and labels are socially and politically constructed.

Key concepts and concerns: language of criminological analysis; subjective elements of crime constitution; media studies

Exemplar: Brisman (2012), who argues for the need to deconstruct categories such as 'crime', 'criminal' and 'victim' in analyses of environmental harm, so that underlying relations of power and the labelling processes can be exposed, as in the case of contrarianism and climate change.

Speciesist criminology

Has a focus on *speciesism* as the main target for criminological research and *critique of anthropocentrism* in the construction of environmental issues, insofar as species and individual members of species are seen to have intrinsic value and rights.

Key concepts and concerns: speciesism as a form of discrimination; abuse of animals, including factory farms; illegal wildlife trade

Exemplar: Beirne (2009), who argues that abuse and degradation of animals has to be analysed in its historical and social contexts, and that major questions need to be answered regarding how, why, where and when animal abuse occurs.

In general, criminologists have often left the study of environmental harm, environmental laws, and environmental regulations to researchers in other disciplines. This has allowed little room for critical examination of individuals or entities that kill, injure, and assault other life forms (human, animal, or plant) by poisoning the earth. In this light, a green criminology is needed to awaken criminologists to the types of major environmental harm and damage that can result from environmental harms; the conflicts that arise from attempts at defining environmental crime and deviance; and the controversies still raging over possible solutions, given extensive environmental regulations already in place.

Typically, there are differences within green criminology around issues pertaining to the distinction between 'harm' and 'crime'. These differences do not stem solely from disputes over the legal/illegal divide, however. There are also profound disagreements with regard to victimisation and varying conceptions of justice. For instance, there may be differences *within* a particular area of work, such as debates over 'animal rights' versus 'animal welfare' in the case of concerns about species justice (Francione, 2010). There are also disagreements in terms of priorities, values, and decision-making *between* particular areas of green criminology (Beirne, 2011; White, 2013). This is evident, for example, in debates over multiple land-use areas. This kind of dispute can involve those who argue that human interests should come first (from the perspective of environmental justice), or that specific ecological niches be protected (from the perspective of ecological justice), even if some animals have to be killed or removed from a specific geographical location. From the point of view of species justice, however, big questions can be asked regarding the intrinsic rights of animals and the duty of humans to provide care and protection for nonhuman species.

Although the unifying link between and among green criminologists is the focus on environmental issues, important theoretical and political differences are nonetheless becoming more apparent over time. For example, some argue that green criminology must necessarily be anti-capitalist and must exhibit a broad radical orientation (Lynch and Stretesky, 2003). Others construe the task as one of conservation and natural resource management, within the definitional limits of existing laws (Herbig and Joubert, 2006; Gibbs et al., 2010a). Still others promote the idea that the direction of research should be global and ecological, and that new concepts need to be developed that will better capture the nature and dynamics of environmental harms in the twenty-first century (White, 2011).

The hallmark of green criminology, regardless of diversity of opinion and the plurality of views, is that proponents argue for more attention to be given to environmental and ecological issues. It is interesting in this respect that a number of prominent criminologists are now utilising their expertise from mainstream areas of criminology (for example, situational crime prevention, general strain theory) to study specifically environmental issues such as illegal trade in elephant tusks, industrial pollution, and social problems arising from climate change (Agnew, 2012; Mesko, Dimitrijevic and Fields, 2010; Lemieux and Clarke, 2009). Green criminology is not only expanding in its own right, but simultaneously there is a greening of criminology more generally.

Differences within green criminology are not only apparent at the level of theoretical focus and orientation. They are also manifest when it comes to responding to environmental crime or harm. For many green criminologists, the biggest threat to environmental rights, ecological justice, and nonhuman animal well-being are system-level structures and pressures that commodify all aspects of social existence, based upon the exploitation of humans, nonhuman animals, and natural resources and that privilege the powerful over the interests of the vast majority. This view is not shared equally among green criminologists, however. In the end, how these questions are addressed has major implications for how responses to environmental harm are framed.

From a critical green criminology view, environmental harm is related to exploitation of both environments and humans by those who control the means of production. Analysis of global capitalism provides answers to questions such as why it is that human societies simultaneously respect and protect certain creatures, especially animal companions like dogs and cats, while allowing and even condoning the dreadful treatment of others, as in the case of factory farming of battery hens to produce eggs (Beirne, 2004; Torres, 2007). It also allows us to better understand why it is that we strive to preserve some environments via creation of national parks, while at the same time devastating particular ecosystems, such as clear-felling of old-growth forests.

Environmental harm takes place within the overarching context of a distinct global political economy. Most writers within the green criminology perspective concentrate on exposing specific types of criminal or harmful environmental actions or omissions. In doing so, they provide detailed descriptions and analyses of phenomena such as the illegal trade of animals, illegal logging, dumping of toxic waste, air pollution, and threats to biodiversity. In many cases, the corpus of work identified within this field has highlighted issues pertaining to social inequality, speciesism, ecological and environmental injustice, and crimes of the powerful. What is less common, however, are examples of study that locate these harms, crimes, injustices, and corrupt practices within the context of an explicit theoretical understanding of the state or economic relations. In other words, it is rare to find a sustained political economy of environmental harm.

Differences in opinion over the nature of global political economy, and over the tactics and strategies most likely to bring about desired social and ecological transformations, manifest in varying approaches to how responses to environmental harm are construed. Thus, there are several ways in which issues pertaining to environmental regulation and the prevention of environmental harm are framed (White, 2008). One approach is to chart up existing environmental legislation and provide a sustained socio-legal analysis of specific breaches of law, the role of environmental law enforcement agencies, and the difficulties and opportunities of using criminal law against environmental offenders. Another approach places emphasis on social regulation as the key mechanism to prevent and curtail environmental harm, including attempts to reform existing systems of production and consumption through a constellation of measures and by bringing nongovernmental and community groups directly into the regulatory process. A third approach presses the need for transnational activism, with an emphasis on fundamental social change. What counts is engagement in strategies that will challenge dominant authority structures and those modes of production that are linked to environmental degradation and destruction, negative transformations of nature, species decline, and threats to biodiversity. Social movements are seen to be vital in dealing with instances of gross environmental harm.

By its very nature, the development of green criminology as a field of sustained research and scholarship will incorporate many different approaches and strategic emphases. For some, the point of academic concern and practical application will be to reform aspects of the present system. Critical analysis, in this context, will consist of thinking of ways to improve existing methods of environmental regulation and perhaps to seek better ways to define and legally entrench the notion of environmental crime. For others, the issues raised above are inextricably linked to the project of social transformation. From this perspective, the focus is on the strategic location and activities of transnational capital, as supported by hegemonic nation-states on a world scale, and how to counter systemic hierarchical inequalities. Such analysis opens the door to identifying the strategic sites for resistance, contestation, and struggle on the part of those fighting for social justice, ecological justice, and animal rights.

There are major political divisions within the broad spectrum of green criminological work (and indeed within green political movements), and these have major implications for whether action is taken in collaboration with capitalist institutions such as corporations and state authorities, or whether it will be directed towards radically challenging these institutions and authorities. Similarly, there are significant tensions between ecological and species justice approaches, as indicated in the following observation:

> The [green environmentalists] rarely champion the sites of their concerns with rights talk, whereas for [animal rights advocates] their very focus is the criterion for moral standing and holding of rights. This crucial deep-seated difference is already present in green criminology in environmentalist notions such as 'fisheries' and 'harvests' and 'conservation', all of which are the stuff and fodder of animal welfare and sustainability but mostly anathema to animal rights. (Beirne, 2011: 354)

To put it differently, some green criminologists view nature instrumentally and harm is viewed through the lens of legality; others view the exploitation of nature, particularly in relation to animals, as intrinsically bad and harmful. How or if this 'moral fissure' can be overcome is of major interest to many currently working under the broad green criminology umbrella.

Conclusion

Two simultaneous processes seem to be taking place in respect to the critical study of environmental crime and harm. The first is the explicit development of 'green criminology' as a distinct and self-conscious area of work within criminology. The second is the more general greening of criminology itself, as more and more mainstream academic researchers undertake research in this area. The expansion of green criminology as a discrete body of work involving particular academics and practitioner networks is, ironically, based upon the notion of exclusivity – that is, that there is something unique and distinctive about this activity called 'green criminology' that sets it apart from other types of social scientific investigation. Conversely, the embrace of climate change and illegal wildlife trade within mainstream criminological circles represents a move toward inclusivity – that is, the field of criminology is sufficiently elastic to allow the incorporation of the study of environmental harm and crime more deeply into its conceptual and methodological universe.

The benefit of labelling this type of scholarly activity as 'green criminology' is that it has provided a focal point for people around the world who share a passionate interest in analysis of, and action around, environmental crime. This has been important in terms of building networks of scholars and researchers and has led to an increasing number and variety of public forums where environmental crime is discussed and debated. Even though it does not preclude individuals from working on their own or in isolation from others, the sense of collective mission has been important in consolidating this area of work, in raising its status and profile within mainstream academic bodies and governmental organisations, and in engendering new conceptualisations and new methodologies. The enhanced circulation and cross-fertilisation of ideas and knowledge has been largely beneficial to all concerned. What unifies the diverse approaches under the green criminology umbrella is a concern with the environment informed by the pursuit of justice, whether this be legal, social, or ecological.

Having said this, it is important to acknowledge the fluid nature of the boundaries and definitions of what is deemed to be 'green criminology'. To illustrate this point, the chapter concludes with descriptions of four different books, all titled *Crimes against Nature* (see Box 1.2).

Perusing the above, we are left with a series of intriguing questions about precisely what it is that we are talking about when we invoke the subject 'green criminology':

- focus on environmental crime?
- focus on environmental harm?
- old methods applied to new area?
- new methods applied to old area?
- subjective choice?
- objective content?
- network and collaborations?

Even though green criminology does denote an established and rapidly growing field of specialisation within criminology – with adherents, entrepreneurs, long-term advocates, occasional participants, and boundary makers – it is, in the end, counterproductive to try to collapse everything environmental into the one box or all writers on environmental crime into one category of intellectuals. The problems of environmental harm persist. The advent of green criminology provides one pathway by which scholars and researchers, individually and collectively, can respond. But the road is wide enough for many fellow travellers.

Box 1.2 Crimes against nature

***Crimes against Nature*, by Robert F. Kennedy, Jr. (2004/2005)**

In this powerful indictment of George W. Bush's White House, environmental attorney Robert F. Kennedy, Jr., charges that the administration has taken corporate favouritism to unprecedented heights – threatening our health, our national security, and our democracy. Kennedy lifts the veil on how the administration, in order to enrich its corporate paymasters, has eviscerated the laws that protect our nation's air, water, public lands, and wildlife. (From the back cover)

***Crimes against Nature*, by Karl Jacoby (2001/2003)**

After European colonialism exported conservation to Africa, Australia, India, and much of the rest of the world, it inevitably spawned new conflicts in these regions as it crossed swords with pre-existing ways of interacting with the environment. What made conservation so controversial in such locales was the fact that it ultimately concerned far more than mere questions of ecology – how many trees to cut and where, what animals to hunt and for how long. In redefining the rules governing the use of the environment, conservation also addressed how the interlocking human and natural communities of a given society were to be organized. (From the introduction)

***Crimes against Nature*, by Donald R. Liddick (2011)**

The alarming consequences of eco-crime go far beyond the widespread degradation of the natural world; important societal institutions are undermined and negative social and economic impacts also result from garbage trafficking, wildlife trafficking, illegal fishing, and illegal logging.... A comprehensive and up-to-date description of each illicit industry is provided, emphasizing the damages caused, the transnational nature of these activities, the roles played by organized crime and public and private elites, and the range of possible solutions. (From the inside flap)

***Crimes against Nature*, by Rob White (2008)**

Environmental issues dominate media headlines and are now having an increasing impact on those who are concerned with crime and criminal justice, leading to the emergence of a 'green criminology'. This is the first book of its kind to provide a comprehensive and coherent overview of green or environmental criminology, as well as charting out new directions for research and thinking in this area. It deals with both the nature of and responses to environmental harm, covering a wide variety of crimes against nature – transgressions against humans, against environments and against animals. (From the back cover)

References

Agnew, R. (2012) 'It's the end of the world as we know it: the advance of climate change from a criminological perspective', in R. White (ed.) *Climate Change from a Criminological Perspective*. New York: Springer.

Beirne, P. (2004) 'From animal abuse to interhuman violence? A critical review of the progression thesis', *Society and Animals*, 12(1): 39–65.

Beirne, P. (2007) 'Animal rights, animal abuse and green criminology', in P. Beirne and N. South (eds) *Issues in Green Criminology: Confronting Harms against Environments, Humanity and Other Animals*. Devon: Willan Publishing.

Beirne, P. (2009) *Confronting Animal Abuse: Law, Criminology, and Human-Animal Relationships*. New York: Rowman & Littlefield Publishers.

Beirne, P. (2011) 'Animal abuse and criminology: introduction to special issue', *Crime, Law and Social Change*, 55: 349–357.

Beirne, P. and South, N. (eds) (2007) *Issues in Green Criminology: Confronting Harms against Environments, Humanity and Other Animals*. Devon: Willan Publishing.

Bell, S. and McGillivray, D. (2008) *Environmental Law* (7th edition). London: Oxford University Press.

Benton, T. (1998) 'Rights and justice on a shared planet: more rights or new relations?' *Theoretical Criminology*, 2(2): 149–175.

Brickey, K. (2008) *Environmental Crime: Law, Policy, Prosecution*. New York: Aspen Publishers.

Brisman, A. (2012) 'The cultural silences of climate change contrarianism', in R. White (ed.) *Climate Change from a Criminological Perspective*. New York: Springer.

Bullard, R. (ed.) (2005) *The Quest for Environmental Justice: Human Rights and the Politics of Pollution*. San Francisco: Sierra Club Books.

Clifford, M. (ed.) (1998) *Environmental Crime: Enforcement, Policy, and Social Responsibility*. Gaithersburg, Maryland: Aspen Publishers.

Francione, G. (2010) 'Abolitionist approach to animal rights', in M. Bekoff (ed.) *Encyclopedia of Animal Rights and Animal Welfare*, Volume 1. Santa Barbara, CA: Greenwood Press.

Fyfe, N. and Reeves, A. (2009) 'The thin green line? Police perceptions of the challenges of policing wildlife crime in Scotland', in R. Mawby and R. Yarwood (eds) *Policing, Rurality and Governance*. Aldershot: Ashgate.

Gibbs, C., Gore, M., McGarrell, E., and Rivers III, L. (2010a) 'Introducing conservation criminology: towards interdisciplinary scholarship on environmental crimes and risks', *British Journal of Criminology*, 50: 124–144.

Gibbs, C., McGarrell, E., and Axelrod, M. (2010b) 'Transnational white-collar crime and risk: lessons from the global trade in electronic waste', *Criminology and Public Policy*, 9(3): 543–560.

Hall, M. (2013) *Victims of Environmental Harm: Rights, Recognition and Redress under National and International Law*. London: Routledge.

Halsey, M. (2004) 'Against "green" criminology', *British Journal of Criminology*, 44(4): 833–853.

Halsey, M. and White, R. (1998) 'Crime, ecophilosophy and environmental harm', *Theoretical Criminology*, 2(3): 345–371.

Herbig, F. and Joubert, S. (2006) 'Criminological semantics: conservation criminology – vision or vagary?' *Acta Criminologica*, 19(3): 88–103.

Jacoby, K. (2003) *Crimes against Nature: Squatters, Poachers, Thieves, and the Hidden History of American Conservation*. Berkeley: University of California Press.

Kennedy, Jr., R. (2005) *Crimes against Nature: How George W. Bush and His Corporate Pals Are Plundering the Country and Hijacking Our Democracy*. New York: Harper Perennial.

Lane, P. (1998) 'Ecofeminism meets criminology', *Theoretical Criminology*, 2(2): 235–248.

Lemieux, A. and Clarke, R. (2009) 'The international ban on ivory sales and its effects on elephant poaching in Africa', *The British Journal of Criminology*, 49(4): 451–471.

Liddick, D. (2011) *Crimes against Nature: Illegal Industries and the Global Environment*. Santa Barbara, CA: Praeger.

Lynch, M. (1990) 'The greening of criminology: a perspective on the 1990s', *The Critical Criminologist*, 2(3): 1–4 and 11–12.

Lynch, M. and Stretesky, P. (2003) 'The meaning of green: contrasting criminological perspectives', *Theoretical Criminology*, 7(2): 217–238.

Massari, M. and Monzini, P. (2004) 'Dirty business in Italy: A case-study of illegal trafficking in hazardous waste', *Global Crime*, 6(3 and 4): 285–304.

Mehta, M. (2009) *In the Public Interest: Landmark Judgement and Orders of the Supreme Court of India on Environment and Human Rights* (volume 1). New Delhi: Prakriti Publications.

Mesko, G., Dimitrijevic, D., and Fields, C. (eds) (2010) *Understanding and Managing Threats to the Environment in South Eastern Europe*. Dordrecht: Springer.

Pearce, F. and Tombs, S. (1998) *Toxic Capitalism: Corporate Crime and the Chemical Industry*. Aldershot: Dartmouth Publishing Company.

Pellow, D. (2007) *Resisting Global Toxics: Transnational Movements for Environmental Justice*. Cambridge: The MIT Press.

Ruggiero, V. (1996) *Organized and Corporate Crime in Europe: Offers That Can't Be Refused*. Aldershot: Dartmouth.

Sankoff, P. and White, S. (eds) (2009) *Animal Law in Australasia: A New Dialogue*. Sydney: The Federation Press.

Shelley, T. and Crow, M. (2009) 'The nature and extent of conservation policing: law enforcement generalists or conservation specialists?' *American Journal of Criminal Justice*, 34(1): 9–27.

Situ, Y. and Emmons, D. (2000) *Environmental Crime: The Criminal Justice System's Role in Protecting the Environment*. Thousand Oaks: Sage.

Sollund, R. (2008) *Global Harms: Ecological Crime and Speciesism*. New York: Nova Science Publishers.

South, N. (1998) 'A green field for criminology? A proposal for a perspective', *Theoretical Criminology*, 2(2): 211–233.

South, N. (2007) 'The "corporate colonisation of nature": bio-prospecting, bio-piracy and the development of green criminology', in P. Beirne and N. South (eds) *Issues in Green Criminology: Confronting Harms against Environments, Humanity and Other Animals*. Devon: Willan.

Torres, B. (2007) *Making a Killing: The Political Economy of Animal Rights*. Oakland, CA: AK Press.

Walters, R. (2011) 'Toxic atmospheres: air pollution, trade and the politics of regulation', *Critical Criminology*, 18: 307–323.

Wellsmith, M. (2010) 'The applicability of crime prevention to problems of environmental harm: a consideration of illicit trade in endangered species', in R. White (ed.) *Global Environmental Harm: Criminological Perspectives*. Devon: Willan Publishing.

White, R. (2008) *Crimes against Nature: Environmental Criminology and Ecological Justice*. Devon: Willan Publishing.

White, R. (2011) *Transnational Environmental Crime: Toward an eco-global criminology*. London: Routledge.

White, R. (2013) *Environmental Harm: An Eco-Justice Perspective*. Bristol: Policy Press.

Williams, C. (1996) 'An environmental victimology', *Social Justice*, 23(4): 16–40.

2

Environmental Victimisation and Conflict Resolution: A Case Study of e-Waste

Lieselot Bisschop and Gudrun Vande Walle

Introduction

Environmental crime and the victimisation it causes is a topic that has been neglected in both victimology and criminology for a long time. In the last decades, criminology has had an increasing focus on environmental crime (White, 2009). This fits within a broader critical development, which looks beyond crime towards legally ambiguous behaviours that cause social harm (Hillyard et al., 2004). The harmful effects of several transnational environmental crimes are the impacts on the quality of water, soil, and air, as well as on the survival of endangered species and on climate change (Walters, 2007; Stretesky and Lynch, 2009). Identifying the victims is not straightforward, because it requires thinking about both geographical and temporal dimensions of victimisation, evoking a more abstract and hidden victim (Goodey, 2005).

Within the field of green criminology, there is attention paid to environmental crime, but not necessarily to environmental victimisation and particular cases of it (Hall, 2011). The debate even seems to neglect the victimisation of individuals and communities and consequently is in danger of disregarding the question of whether these victims have a right to and the possibility of conflict resolution. In the literature on corporate crime, there is ample discussion of regulation and prevention, and also discussion related to environmental issues, but research on victimisation by corporate crime is limited (Friedrichs, 2002; Croall, 2007; Vande Walle, 2012). A few notable exceptions have focused on harms caused by crimes of the powerful or activities both 'lawful and awful' (Box, 1983; Pearce and Tombs, 1998; Passas and Goodwin, 2004).

Generally speaking, criminology thus has kept rather silent about environmental crime victims (Skinnider, 2011).

The little sister of criminology called victimology has made important steps in the improvement of the position of the victim within both research and policy. The focus, however, has been on the rights of victims, rather than on victims as objects of harm (White, 2011). At the origin of victimology, Mendelsohn (in Drapkin and Viano, 1974) defined victims according to their contribution to the crime. Later, a broader variety of victims was considered, such as victims of work accidents and genocide, and even victims of events beyond human control, such as natural disasters (Viano, cited in Elias, 1986). This was a promising first step for environmental victimology. Today, it seems victimology has taken a step backward. Through its link with criminology, it has become set on defining the victim in relation to the criminal justice system. Hidden crimes – hidden to law enforcement, that is – seem to stay off the victimology radar. Victimology tends to predominantly focus on traditional crimes and on individual rather than collective harms. From 2000 onwards, this traditional interpretation has changed in favour of discussing mass victimisation in postconflict areas, although the debate is still largely focused on political cases (Stover and Weinstein, 2004; Aertsen et al., 2008).[1] Therefore, during the 2012 'World Conference of Victimology' in the Hague, the session on environmental victims organised by Matthew Hall was welcomed as a new topic.

In this chapter, we explore victimisation of transnational environmental crimes and possibilities for conflict resolution. This is not limited to strict legal definitions of crime or victimisation: environmental victims are those harmed by changes in their environment due to deliberate or reckless acts or omissions (Williams, 1996). We have selected one case: the victimisation by illegal e-waste (electronic waste) transports and the transports of secondhand electronics. often resulting in dismantling, burning. and dumping of hazardous components in developing countries. These transports from European harbours to the dump sites in Ghana have characteristics that may hinder prevention and may complicate thinking about victimisation and redress (Bisschop, 2012). This chapter intends to provide insights about the victimisation and explore possibilities for conflict resolution.

Method

The following analysis is based on a case study research of illegal transports of e-waste within the research setting of the Port of Antwerp and its consequent transport to and dumping in Ghana. The document analysis is based on governmental sources, research reports, corporate documents, and reports by nongovernmental organisations (NGOs) and the media. A total of 56 semi-structured interviews were conducted with representatives of government agencies,[2] corporations[3] as well as representatives of civil society.[4] These respondents[5] were located within the Belgian research setting and other European Union (EU) countries as locations of origin and in Ghana as one of the countries of destination. This research included field visits as well, limited to crucial sites and actors for the e-waste case. For this chapter, the field visit in Ghana was particularly relevant. As a country of destination, Ghana is often frequented by illegal transports of e-waste that export from or transit in Belgium. In Ghana, we studied the port of Tema, the informal recycling and refurbishing firms and e-goods markets in the cities of Tema and Accra, and the Agbogbloshie (a suburb of Accra) dump site. We coded and analysed data gathered in both desk research and interviews,[6] making it possible to triangulate findings (King, Keohane and Verba, 1994; Yin, 2009).

Scope, orientation and actors involved in e-waste

Measuring illegal transports of e-waste: 'best guesstimates'

There is little official data assessing the scope of illegal transports of e-waste. The existing reporting systems are generally mediocre (Fischer et al., 2008) because of a lack of consensus about the definition of the goods codes and because many countries, including those within the EU, do not report. Even the statistics that are provided are not necessarily accurate reflections of the actual illegal transports occurring, but rather of the law enforcement prioritisation of the issue, or lack thereof. Moreover, trade statistics often do not include the data on secondhand electronics, a major consumption flow from corporate and government consumers (VROM-inspectie, 2011). Available data are thus 'best guesstimates', approximations of the real scale. This is particularly worrying because it has implications for the recognition of victimisation in both literature and policy making (Gibbs and Simpson, 2009). Uncertainty about the scale of crime undermines the reliability of the data and influences the perception of the seriousness of its victimisation (Slapper and Tombs, 1999).

The amount of waste of electrical and electronic equipment (WEEE) generated is estimated to grow 3 to 5 per cent each year, adding up to 15 to 20 kg (33 to 44 lbs) of e-goods brought onto the market per capita per annum (Crem, 2008). About 7 per cent of this is registered as WEEE exports and up to 33 per cent is separately collected for environmentally sound treatment (Environmental Investigation Agency, 2011). The European Commission has noted that 'A part of the other two thirds is potentially still going to landfills and to sub-standard treatment sites in or outside the European Union' (European Commission, 2011). Estimations of the global scope of illegal e-waste flows vary from 20 to 25 million tonnes (22 to 27.5 million tons) per year (Robinson, 2009) to 20 to 50 million tonnes (22 to 51 million tons) per year (United Nations Environment Programme, 2006). Many of these transports are wrongly declared secondhand goods or metal scrap, although they contained chlorofluorocarbons (CFCs)[7] or cathode ray tube (CRT)[8] monitors (Heiss et al., 2011). A Ghanaian terminal operator revealed he has several clients that ship approximately 20 containers of WEEE or used electronic and electric equipment (UEEE) per month. He was just one of several terminal operators in one of the several ports in the West African region.

Orientation of the flows

Reports indicate that the flows of WEEE/UEEE go from Western Europe and the United States to West Africa and Southeast Asia (Crem, 2008; Fischer et al., 2008). Within Europe, the countries with the biggest harbours (the Netherlands, Germany, Belgium, and the United Kingdom), major transit locations for inland Europe, are those with the biggest share of transports of e-waste, both legal and illegal (IMPEL-TFS, 2005; de Rijck, 2011). These ports are also those investing most in controls for illegal transports of e-waste, increasing their statistics on illegal trade, which in turn, give them a reputation of being illegal waste hubs. Other ports do less controls and thus have lower numbers of illegal trade detected. This makes the former victim to their own law enforcement success (van Erp and Huisman, 2010).

In today's globalised economy, the asymmetries in legislation, knowledge, economy, and culture shape the illegal flows of waste and e-waste (Passas, 1999; Gilbert and Russell, 2002). Asymmetries in environmental regulation and enforcement as well as in prices for recycling and disposal have led shippers to go in search of the cheapest 'recycling' solution. Because of their economic and trade dependence, developing countries often allow these imports. Importers of electronics pay import taxes for each new or second hand item they bring into Ghana. This generates tax revenue for the country. Other asymmetries that shape this are those in knowledge and awareness. The field visits made it clear to us that many people, both ordinary citizens and law enforcement people, are unaware of the harmfulness of e-waste and do not have the knowledge or facilities to dismantle and recycle the equipment properly. These various asymmetries are at the basis of the differential victimisation.

Actors involved along the supply chain

Actors throughout the trade flow have the potential to shape illegal transports of e-waste. Earlier publications (Gibbs, McGarrell and Axelrod, 2010) referred to profit or lure as the major aetiological factors, but several other elements play a role in the emergence of illegal e-waste flows and the consequent 'recycling' – dumping and burning. In exporting and importing countries and in supply and demand, motivations and opportunities are ample. Illegal e-waste flows start at the consumers, who seek a way to get rid of their no longer functioning television sets, computers, and so on, and extend to e-waste collectors, recyclers, and refurbishers in developed as well as developing countries (Schluep et al., 2008; Bisschop, 2012). The nature and scope of the collectors, recyclers, and refurbishers is impossible to grasp without taking the broader context into account (van Erp and Huisman, 2010).

E-waste is one of the fastest-growing waste markets and is likely to increase in the coming years, given the exponential consumption of electronics (UNEP, 2005; Pellow, 2007). Both producers and consumers have a responsibility in this. Producers can ensure that the recycling of e-waste is less harmful by phasing out hazardous components and through eco-design.[9] Consumers have the responsibility of ecological consumption. Once discarded, e-waste or secondhand electronics can end up in illegal transports, with countries like Ghana as final destinations. Exporting the waste is a way to externalise the harm and create a distance between producers and consumers, on the one hand, and those affected by the dumping or recycling of the products, on the other hand (White, 2011). Many of the e-waste collectors and recyclers live up to their espoused environmental and ethical standards and regard illegal transporters of e-waste as their biggest competitors. Other organisations that claim to recycle WEEE/UEEE are less honourable and engage in export – often through brokers – to developing countries.

Besides actors in countries of origin and transit, the countries of destination play a role in attracting the flows of e-waste or secondhand electronics. The institutional framework in countries of destination – or lack thereof – contributes to the flows (Schluep et al., 2008). Although Ghana, like many other West African countries, has signed the Basel Convention as well as other international conventions about guaranteeing a healthy and safe environment for its citizens, the transposing that desire into national legislation has been lacking. A law about criminalising illegal e-waste transports has been prepared by policy makers (for example, the Ghanaian Environmental Protection Agency and the Ministry of the Environment), but it has been difficult to get this through Parliament. Several issues play a role in this. First of all, through taxes, the imports of secondhand electronics are a major source of revenue for these countries (Basel Action Network and Coalition, 2002; Pellow, 2007). Second, the digital divide creates a hunger for technology in developing countries. Computers, mobile phones, and other electronic devices allow people to catch up with global developments in knowledge and communication. Third, the e-goods discarded by industrialised countries provide a sole secure source of livelihood for many people in developing countries. Secondhand – but also nonworking – television sets, computers, mobile phones (or their batteries), and so on are sold on many street corners. These economies in transition and developing countries have a massive formal as well as informal economy thriving on repair, refurbishment, dismantling, and recycling (Prakash and Manhart, 2010; Amoyaw-Osei et al., 2011). The collection and dismantling system in Accra is very effective and guarantees the inflow of metals into recycling facilities. The informal e-waste collectors and 'recyclers' feed into the legal e-waste industry through the increased demand for secondary raw materials. These dismantlers are encouraged to sell the motherboards and other valuable components for recycling. This of course exposes the workers to the risk of being exploited for their cheap labour. Currently, some producers are organising the take-back of discarded electronics, but several Ghanaians experienced this to be 'cherry-picking': taking back only the valuable components (for example motherboards). There are no details available about the final destination of the components or metal scrap, but in Ghana the motherboards of the dismantled computers were found to be sold whole for export to Nigeria or China or, as they told us, they sell these to 'white men from Europe'. Other recovered metals are used in local industries or sold for export (Prakash and Manhart, 2010). Local NGOs in Ghana teach dismantlers how to extract the precious metals without burning, but because burnt copper cables are still a valuable good – often bought by Asian buyers – the burning continues.

E-waste and its victims

A major share of the EEE that is transported to developing countries never makes it to the secondhand market, but is dismantled to extract the raw materials. All too often this 'recycling' happens under precarious circumstances, where remainders are illegally dumped or burned, releasing the toxic components[10] (Eidgenössische Materialprüfungs- und Forschungsanstalt [EMPA], 2009). These toxins easily disseminate in the soil, air, and water, often without notice, but they are not easily broken down and thus stay there for many years (Baker et al., 2004; Greenpeace, 2008a; European Environment Agency, 2009). In reference to the butterfly effect (White, 2011), the harm manifests itself locally, regionally, nationally, and globally.

The first environmental harm is the impact of the hazardous components upon ecosystems. A second environmental effect is on climate change, through the release of the greenhouse gases they contain, and especially through the burning of products. For one device the impact is minimal, but the mass quantity of batteries, monitors, and so on makes it problematic (Leefmilieu, natuur en energie [LNE], 2010). Moreover, the continued exposure to these hazardous substances can make the impact much greater than initially experienced (South, 1998). In merely referring to the effects on the quality of water, air, and soil, we risk ignoring the human victims of environmental harm. Speaking in terms of the impact on ecosystems and climate change risks considering the phenomenon as 'victimless', a statement so often related to corporate crime. This seems to 'exempt' the responsible actors from caring for the victims or remedying the harm (Fattah, 2010). Victims of the e-waste dumping often do not know they are harmed, or they accept the harm because they need the e-waste business to survive. Environmental victimisation is not always immediately visible or identifiable, but victims of environmental crime and harm do exist. Naming these victims and discussing redress is important (Hall, 2011).

The improper dismantling and recycling of e-waste has a detectable impact on the health of workers. The precarious working circumstances in recycling facilities in Africa and Asia in terms of health, safety, and working standards have been illustrated multiple times (Basel Action Network and Coalition, 2002; Basel Action Network, 2005; Greenpeace, 2008a, b; Sepúlveda et al., 2010). In the West African countries of Ghana, Togo, Nigeria, and Ivory Coast, adults and children go through the heaps of dumped electronic and electrical equipment in search of valuable materials, often barehanded, and dismantle them without protective equipment. Asia seems to have a somewhat better reputation with recently installed treatment facilities, but a corporate respondent (C15) said, 'Often the façade in Asia is good, but you should not look too far inland'. A Swedish study found that e-waste flows are moving to the north of China or to Vietnam and Cambodia, likely as a result of more stringent environmental policy in Southern China (Nordbrand, 2009). E-waste workers in developing countries, whether in Africa or

Southeast Asia, generally do not have the cultural nor the economic capital to adapt to the environmental harm they are faced with (South, 1998). As an NGO respondent explained, many of the e-waste workers, often minors, work at Agbogbloshie, a suburb of Accra, Ghana, for 3 to 5 years, after which they return to their families in the north, because they can no longer work due to their illnesses, basically returning home to die. Children are extremely vulnerable to all kinds of deteriorating environmental conditions (Stephens, 2009).

Not only those working in the e-waste business are affected by the harmful substances. People living nearby, who use the water contaminated by the dumping or burning and inhale the toxic smoke of the burning e-waste, and as well as those eating the crops grown on the toxic soil, feel the effects.[11] CRT monitors, for example, contain lead, which is known to damage the nervous, kidney, blood, and reproductive systems and affect children's brain development. The monitors contain barium as well, causing brain swelling, heart damage, and increased blood pressure (Basel Action Network and Coalition, 2002). The gathering of soil and water samples in longitudinal studies is pertinent to demonstrate this victimisation. Through the contaminated water, air, and soil, environmental victimisation travels across local and national borders. Think for instance of the fruits and vegetables we import from regions that import e-waste. Although the quantities are likely to be minimal, the Northern 'not in my backyard' viewpoint backfires in these instances.

An important element to take into account here is the link of illegal e-waste transports to social and ecological inequalities. Vulnerable groups are also those most likely to suffer, because they have a bigger chance of working at or living near the polluting factories or landfills (Stretesky and Lynch, 1998). Their social, economic, and political characteristics make them vulnerable to victimisation, resulting in e-waste flowing to the poorest nations (Pellow, 2007; White, 2011). This differential victimisation occurs because of differences in cultural capital and in information about the harmful effects of working and living in a highly polluted environment. Several Ghanaian respondents explained that importers and sellers of secondhand goods constitute an important group of voters, causing politicians to hesitate to draft more stringent regulations. Moreover, those most severely harmed by the burning and dumping – to be precise, those informal workers at the Agbogbloshie dump site – constitute a minority group who travelled to Accra from the north of Ghana in search of an alternative sources of income, which their region could no longer provide. They are also a religious minority and therefore not the biggest concern for politicians in Accra. This is to a certain extent a deliberate neglect of the environmental victims and can even be called environmental racism (Stretesky and Lynch, 1998).

Economically and politically, these illegal transports of e-waste can also be harmful. The cleanup of dumps is a heavy burden for developing countries. Illegal transports also have economic advantages over legitimate transports, because of lower processing costs, and therefore they adversely affect trade and competition. Facilities that do work in an environmentally friendly way when collecting and recycling experience these illegal transports as false competition. Moreover, the informal 'recycling' sector has a lower recovery rate of precious metals, which negatively impacts the availability of natural resources. Politically, this illegal trade also undermines the often already weak law enforcement of developing countries through corruption and fraud, and it mocks international policy making (Quadri, 2010). Finally, there is victimisation of Belgian and other European consumers, because upon buying new electronics they also pay for recycling.

The challenges of e-waste victimisation

The complexity of the e-waste case is a challenge to the discussion about realistic ways of conflict resolution among the responsible actors and the victims. A first element to consider is the normalisation effect of this continuous dumping and dismantling of e-waste, as opposed to one-time, one-place environmental disasters. The cases of Chernobyl, Bhopal, Deepwater Horizon,[12] and Fukushima each had a major impact, causing international outrage. Because these disasters are rather clear cases with identifiable actors, we expect that a responsible actor will be named. Even in those cases, however, such actors are often not held criminally accountable. The trial[13] of Trafigura for the illegal dumping of hazardous waste from the *Probo Koala* ship was a rare exception. Compared to these major environmental cases, the dumping of e-waste seems to be a more 'silent' environmental harm, in line with overfishing or illegal logging. These phenomena affect many people day after day, silently, often with a late onset of harm, only sporadically attracting the attention of the news media, NGOs, or researchers.

Second, and related to the normalisation effect, is the idea that victims participate in their own victimisation. Terms such as victimity and victim precipitation are part of the history of organisational criminology and victimology and reveal an endemic idea that victims are to a certain extent responsible for their victimisation (Mendelsohn in von Hentig, 1948; Drapkin and Viano, 1974). Blaming the victim was also prominent in the Union Carbide case (Bhopal) where the Indian employees were blamed for not caring for the safety of the facility (Pearce and Tombs, 1998). Applying this to the e-waste case, where the workers 'choose' to work at Agbogbloshie, does that mean they can no longer be perceived as victims of environmental crime? Should the extent to which they are considered victims depend on the extent to which we can hold them accountable for working or living there? This blaming of the victim, whatever its degree, holds the risk of downplaying the harm.

A third challenge is the blurred boundary between corporate crime, organised crime, and the informal economy. Illegal transports of e-waste have characteristics of each, making it difficult to identify the responsible actors. Throughout the supply chain of electronics, several actors can feed into illegal transport and dumping of e-waste. Some merely facilitate illegal activities, whereas others shape the flows through more deliberate actions and omissions. Corporate actors are involved in production, collection, transport, and recycling, and flows of e-waste and secondhand goods have a major share in the Ghanaian economy. Therefore, there is a risk of governance activities merely focusing on the prevention and control of the trade instead of taking into account the health problems of the people working on and living near the dump sites. Some local governments in Belgium even fear corporate bankruptcy and job loss if controls on e-waste transports are too severe. The same applies in Ghana, where the economic pressure of major corporations put this country in a weak negotiating position. The focus of regulation is in danger of being the protection of state income rather than concern for the victim.

Sometimes the connection with organised crime syndicates is mentioned (Interpol, 2009). During the field visits in Ghana, we found that the buying of metal scrap on the waste dumps was considered to be in the hands of organised crime. Syndicates of Nigerian, Italian, Eastern European, and Chinese origin buy the valuable materials from informal workers on the dump site and sell them on the global metal market as secondary raw materials. The aspect of organised crime and its interplay with the informal economy again relates back to the victims participating in the activity under dirty, dangerous, and difficult circumstances, but nevertheless making a living. Believing that this is merely the victims' job might be a neutralisation technique for many Western producers and consumers. The large distance between the producers and consumers and the dump sites might explain the lack of awareness that consumers have about the harmfulness of their e-goods consumption. Recycling the products in environmentally effective facilities – for example, in Belgium – is expensive. Because countries do not want the recycling to be done in their backyard (the NIMBY syndrome), a solution is sought in illegal export. In this way, developed countries choose the environmental victimisation of developing countries rather than their own economic harm.

A final element to consider is the competition of victims (South, 2007; White, 2007). Should the priority be with the environment as a victim, or with the Ghanaian informal workers, or with the consumers or European industry, which loses valuable secondary raw materials? This competition among social, environmental, and economic victimisation makes it difficult to determine the focus to redress.

Discussing restorative justice and conflict resolution possibilities

The above-mentioned characteristics make thinking about conflict resolution complex. Nowadays, victims are generally acknowledged in justice procedures (Goodey, 2005). The idea of restorative justice has inspired many judicial institutions. This recognition of victims and their rights raises the question of how to redress the harm. Do victims merely earn financial compensation or do they get a say in the criminal justice procedure? For environmental harm, the first concern is how to stop the destruction and prevent further pollution. A second concern is the issue of reparation or compensation. This second concern, however, often remains an issue in the margin. A few prominent cases have been studied by researchers in organisation al and green criminology. The Union Carbide gas leak in Bhopal, India; the *Probo Koala* case; the international asbestos cases; and the Fukushima nuclear plant aftermath are laboratories for conflict resolution and resistance against it. The picture is different for the e-waste case: the responsible actors cannot easily be singled out and are often nowhere near the affected area; the media reporting is limited; and the value for the local economy keeps victims from requesting redress. Together these elements ask for a more fundamental study of conflict resolution of environmental crime that goes beyond the single case study.

In aiming to contribute to this fundamental study of environmental victimisation and conflict resolution, we bring in the aspect of restorative justice. This pays attention to the prevention of future harm without neutralising or denying the harm inflicted (Aertsen, Vanfraechem and Crawford, 2010). It aims for the reintegration of offenders and the reparation of the harm imposed (Braithwaite, 2002). It intends to discuss emotional and psychological dimensions of harm (Strang and Sherman, 2003). Restorative justice requires involving all the parties: the victims, the community, and the actors responsible for the harm. Discussing environmental victimisation then requires moving away from the traditional dichotomy between victim and offender (McEvoy and McConnachie, 2012). The e-waste case poses a challenge for restorative justice: How do we take into account present-day and potential future harm? How do we determine who the victims are? How do we name the responsible actors and make them recognise their responsibility? Finally, which authority will support a restorative justice initiative?

Restorative justice by victim-oriented and community-oriented prevention

From 2000 onwards, restorative justice studies focuses on crime prevention as a fundamental element for restoring harm. As Cavadino mentions: 'We should cease to look to severity of punishment to control crime... [but] should look primarily to measures of crime prevention outside the criminal justice system' (Cavadino, 1999, cited in Aertsen et al., 2010). This is certainly the case for environmental crime and its harms.

In Ghana, as in other countries of destination of e-waste, a first priority is with addressing the structural causes. This refers to the need for a source of livelihood, the desire to bridge the digital divide, the demand for raw materials, and the lack of recycling facilities. Small steps have been taken to address these issues. Capacity-building projects have been set up to raise awareness and to look for solutions to deal with the immediate harm. In Accra, the initiatives involve teaching the informal actors how to improve their working and living conditions with simple tools to dismantle rather than burn the e-waste. These projects have been shaped by local NGOs, often with the financial support of international governmental or other nonprofit institutions. Internationally, it is also mainly NGOs that have advocated for these victims and tried to remedy their environmental victimisation. In the 1970s and 1980s, the media reporting and NGO actions resulted in the drafting of international conventions. Today, the naming and shaming activities[14] again bring the topic to the attention of corporations, consumers, policy makers, and law enforcement. Similar to fighting corporate crime, it is thus mainly NGOs that advocate for environmental victims (Croall, 2007). On the community level, they take initiatives to avoid the immediate effects of dumping and burning e-waste in developing countries. On an individual level, they try to convince consumers about the hazardous effects of nonsustainable consumption. On a macro level, they try to make people aware of the consumption and production patterns in today's society, the treadmill that provides the fertile breeding ground for illegal transports and dumping of e-waste (Long et al., 2012).

It can prove useful to involve a broader network of actors in preventing illegal transports from occurring. This requires corporate actors to cooperate with NGOs and governments. For instance, increasing the awareness of transport actors through due diligence requirements in European legislation could be explored. Involving corporate actors also has the advantage of being able to use their expertise to better inform risk assessments in view of controls (for example, in countries of origin). Networked governance is severely challenged when goals, means, and responsibilities differ (Bisschop, 2013).

These preventive measures are indispensable to reduce the risk of future harm and to take into consideration the economic benefits of e-waste. The question of restoring or compensating the harm to individual victims and to the community remains unanswered.

Retribution as a final step

The debate about conflict resolution in cases of organisation al crime was long dominated by the need for a criminal justice procedure. Braithwaite mitigated this strong belief in punitive measures: criminal justice is just an instrument to punish, at the top of the pyramid, when all the other instruments on the levels below have failed (Ayres and Braithwaite, 1992; Braithwaite, 2008). Criminal justice responses to environmental crime are rare. These cases are usually settled by inspectorates and consequently dealt with through compliance (Gunningham, Kagan and Thornton, 2003). Despite illegal transports of e-waste being criminalised, an important challenge remains the underfunding of these enforcement agencies with consequences for training, resources, and effective follow-up throughout the flows (Brack and Hayman, 2002). Illegal e-waste flows are inherently transnational, but much of the implementation remains local and fragmented. Law enforcement is perceived to be very slow in response and inadequate in determining the fines.

Due to legal asymmetries and demanding evidence gathering, it is difficult to prosecute the actors who are responsible. Cardwell, French and Hall (2011) have, however, stressed the importance of the criminal law for the recognition of environmental victims. They explored Directive 2008/99 EC of the European Parliament on protection of the environment through criminal law and asked how far this directive may help in defining environmental crime victims. This requires EU member states to criminalise violations of the EU environmental legislation (for example, the illegal shipment of hazardous waste). The authors problematise the lack of connection to victim rights and more in particular that of victim participation. Cardwell et al. (2011) refer to the UN *Declaration of Basic Principles of Justice for Victims of Crime and Abuse of Power,* which affords victims access to justice, fair treatment, compassion, and respect for the victim's dignity, the right to redress, restitution, and compensation. In another study, Huisman (2009) explored the applicability of the Rome Statute, which defines the field of application of the International Criminal Court (ICC). He wonders if the ICC could prosecute companies for committing transnational corporate crime that has disastrous effects for humanity. The criminal law procedure on an international level could then be considered a way of punishing the actors of transnational environmental harm, a domain that merits further exploration. The criminal justice system is an indispensable part of the pyramid. However, we have to bear in mind that, except for a few countries, criminal justice systems are not generally made for compensation or reparation of harm. The criminal conviction of a polluting company can be a relief for victims, because the responsibility has officially been recognised (Croall, 2001). It is not an official means for conflict resolution.

We have discussed two restorative justice mechanisms: victim- and community-oriented prevention and criminal justice. Other mechanisms were only named: out-of-court settlement, law enforcement by the inspectorate, naming and shaming. We ask for the fundamental and structured study of environmental victimisation and restorative justice for victims of transnational environmental crime and harm. It is a challenge for green criminology to study in how far the existing mechanisms of redress respond to the conditions of restorative justice. Do victims appreciate the out-of-court settlements as a step toward restorative justice? What participation, recognition, and redress do they expect?

Conclusion

We have tried to illustrate the victimisation related to illegal transport and dumping of e-waste. Although the transport of e-waste to developing nations has been criminalised, the legal transport of secondhand electronics can have equally detrimental effects, because these countries do not have the necessary recycling facilities. In studying environmental victimisation, it is therefore important to use a scope broader than the mere legal definitions and include 'less visible' harms. This case study also illustrated how environmental victimisation is inextricably connected to the everyday functioning of our society and 'is not a socially neutral process' (White, 2011).

The victimisation of environmental crime is complex. It may have been occurring for several years, often decades or longer, hence the complicated question of redress. This is evident for the e-waste case, but similarly applies to other types of environmental harm. Can we, for instance, determine who has to pay the costs of adapting to climate change or mitigating its consequences? In studying and addressing environmental victimisation, we deal with nontraditional victims, those generally left out of victim surveys. Harmed individuals, communities, and ecosystems are less visible, but nonetheless are victims. Environmental crime challenges our traditional view of the victim. The case of e-waste illustrates how these can be made visible simply by following the flows to their eventual destination, although the temporal and geographical scope is much more difficult to assess. Dealing with this complexity through the criminal justice system is difficult, because it is traditionally focused on one-time, one-place events. Nevertheless, studying environmental victimisation and raising awareness can result in criminal justice systems being more inclusive towards these victims (Croall, 2007). Moreover, it allows us to highlight other means to address environmental victimisation. The risk of environmental harm needs to be reduced throughout the different levels that shape it. This implies involving several stakeholders, while taking into account that their perceptions on both harm and solutions might differ (Bisschop, 2013).

Environmental victimology has a role to play in the prevention of further environmental victimisation and in repairing harm along the principles of restorative justice. In order to stay on top of or at least in pursuit of what is happening in the world of environmental crime, research in both criminology and victimology needs to be able to answer to the changed context of victimisation by environmental crime in a globalised world. The victim, whether human, animal, or eco-system, is not necessarily located in our own backyard. Studying environmental victimisation and conflict resolution requires researchers to demonstrate criminological imagination to cope with the challenges of 'unknown' victims and responsible actors within the complexity of the transnational context. This undoubtedly requires us to look beyond the limitations of our own discipline and its definitions, but similarly requires us to embrace the existing knowledge and apply it to 'new' areas of concern such as environmental victimisation.

A last but important remark is the idea of colonising people by implying the status of victim. We unavoidably wrote this chapter starting from our viewpoint, based in a European country, and focused on the harmful effects of e-waste dumping. Are we right to label the people from Tema and Accra the victims, or is that a patronising and Western point of view? Is that also the way the Agbogbloshie workers and residents see themselves? Do they, for instance, expect conflict resolution in court or based on our principles of the justice system? Restorative justice has been a globalising force as much as traditional Western legal forms of retribution and rehabilitation have been (Cunneen, 2002). It is thus important to give these local stakeholders a role in environmental governance (Holley, Gunningham and Shearing, 2012), even for complex phenomena. This can only be explored by going to the sites and talking to people, which we have initiated in this research project. This requires collaboration with NGOs, interest groups, and citizens living in the affected communities. This will further the development of an environmental victimology, away from the often fragmentary thinking about victims of environmental crime, hopefully contributing to the prevention of environmental victimisation.

Notes

1. See, for instance, the September 2012 issue of the *European Journal of Criminology*, which is a special issue on Atrocity Crimes and Transitional Justice and pays particular attention to mass victimisation (Vol. 9, No. 5).
2. The governmental actors in this research are national and international government agencies such as customs, environmental inspectorates, police organisations, and prosecution services and administrations.
3. The corporate representatives who were interviewed are producers of computer hardware, e-waste collectors, refurbishers and recyclers, and transport corporations.
4. The civil society respondents in this research are environmental NGOs, union representatives, and research journalists.
5. These respondents were guaranteed anonymity and therefore we refer to government (G and number), corporate (C and number), and civil society respondents (S and number) in quotations.
6. NVivo qualitative data analysis software, QSR International Pty Ltd. Version 8, 2008.
7. Chlorofluorocarbons. CFCs were used as refrigerants (for example R11, R12), propellants, and solvents; but have been phased out by the Montreal Convention because they contribute to ozone depletion.
8. Cathode ray tube, main component of television sets until around 2010. Television sets with cathode ray tubes contain lead (up to 2.5 kg or 5.5 lbs), mercury, and cadmium, components that are hazardous to the environment.
9. This is the subject of the EU's Restriction of Hazardous Substances (RoHS) Directive (2002/95/EC).

10. For example lead, cadmium, brominated flame retardants, beryllium, and mercury.
11. See also: 'The impacts of electronic waste disposal on the environment and public health in the occupied Palestinian territory: A case study from Idhna Hebron Governorate', Applied Research Institute Jeruzalem (ARIJ) (2012). 'Water and Environment Research Department', in Cooperation with Sunflower Association for Human and Environment Protection (http://www.arij.org/files/).
12. News media reported that BP agreed to pay record criminal penalties for the US oil spill http://www.reuters.com/article/2012/11/16/us-bp-spill-idUSBRE8AE1AC20121116 (consulted 18 November 2012).
13. In 2006 in Abidjan, Ivory Coast, 600 tons (544 tonnes) of toxic material was dumped on waste sites near the city. They were transported from Amsterdam with the *Probo Koala* tanker, chartered by a Dutch company Trafigura. According to the UN Special Rapporteur on Toxic Waste, 15 people died, 69 people were hospitalised, and over 100,000 needed medical attention as a result. Trafigura was convicted by the Dutch criminal court and fined €1 billion.
14. For example, the publishing of tags found on dumped e-waste or through confronting previous owners with their old data found on the carelessly discarded e-waste.

References

Aertsen, I., Vanfraechem, I. and Crawford, A. (2010) *Restorative Justice and Crime Prevention. Presenting a Theoretical Exploration, an Empirical Analysis and the Policy Perspective.* Final report of the European project *'Restorative Justice and Crime Prevention'*. Italy: Department of Juvenile Justice. Available at http:/euforumrj.org/assets/upload/Restorative_Justice_and_Crime_Prevention_Final_report.pdf

Aertsen, I., Vanspauwen, K., Arsovska, J., Rochne, H. and Valinas M. (eds) (2008) *Restoring Justice after Large-Scale Violent Conflicts Devon.* Devon: Willan Publishing.

Amoyaw-Osei, Y., Opoku Agyekum, O., Pwamang, J. A., Mueller, E., Fasko, R. and Schluep M. (2011) *Ghana e-Waste Country Assessment. SBC e-Waste Africa Project,* Green Advocacy Ghana (Green Ad), Environmental Protection Agency (EPA), Secretariat of the Basel Convention (SBC), United Nations Environment Programme (UNEP), Eidgenössische Materialprüfungs- und Forschungsanstalt (EMPA). Available at http://wscep.org/Posts/581/Documents/Amoyaw-Osei_2011_GreenAd-Empa.pdf

Ayres, I. and Braithwaite, J. (1992) *Responsive Regulation: Transcending the Deregulation Debate.* New York: Oxford University Press.

Baker, E., Bournay, E., Harayama, A. and Rekacewics, P. (2004) *Vital Waste Graphics,* Basel Convention, GRID, UNEP and DEWA Europe.

Basel Action Network (2005) *The Digital Dump. Exporting Re-use and Abuse to Africa.* Seattle, WA, The Basel Action Network (BAN).

Basel Action Network and S. V. T. Coalition (2002) *Exporting Harm: The High-Tech Trashing of Asia.* Seattle, WA: The Basel Action Network (BAN); San Jose, CA: Silicon Valley Toxics Coalition (SVTC).

Bisschop, L. (2012) 'Is it all going to waste? Illegal transports of e-waste in a European trade hub', *Crime, Law and Social Change*, 58(3): 221–249.
Bisschop, L. (2013). 'Go with the e-waste flows. The governance reality of illegal transports of e-waste in a European trade hub', in P. van Duyne, J. Harvey, G. Antonopoulos, K. von Lampe, A. Maljević and J. Spencer (eds) *Human Dimensions in Organised Crime, Money Laundering and Corruption*. Nijmegen: Wolf Legal Publishers.
Box, S. (1983) *Power, Crime and Demystification*. London: Tavistock.
Brack, D. and Hayman, G. (2002) International Environmental Crime: The Nature and Control of Environmental Black Markets. *Sustainable Development Programme, RIIA workshop, 27–28 May 2002*. London: Royal Institute of International Affairs.
Braithwaite, J. (2002) *Restorative Justice and Responsive Regulation*. New York: Oxford University Press.
Braithwaite, J. (2008) *Regulatory Capitalism: How it Works, Ideas for Making it Work Better*. Cheltenham: Edward Elgar.
Cardwell, P., French, D. and Hall, M. (2011) 'Tackling environmental crime in the European Union: the case of the missing victim?' *Environmental Law and Management*, 23(3): 113–121.
Crem (2008) *Een analyse van stromen electronica-afval in Nederland* [Analysis of e-waste flows in the Netherlands]. Amsterdam: Crem Milieu Onderzoeks- en Adviesbureau.
Croall, H. (2001). *Understanding White Collar Crime*. Buckingham, Philadelphia: Open University Press.
Croall, H. (2007) 'Victims of white collar and corporate crime', in P. Davies, P. Francis and C. Greer (eds) *Victims, Crime and Society*. London: Sage Publications, pp. 78–108.
Cunneen, C. (2002) 'Restorative justice and the politics of decolonization', in E. Weitekamp and H. Kerner (eds) *Restorative Justice: Theoretical Foundations*. Devon: Willan Publishing, pp. 32–49.
de Rijck, R. (2011) A flaw in the criminal approach of international waste transport in Europe. *INECE 9th International Conference*. British Columbia: Canada.
Drapkin, I. and Viano, E. (1974). *Victimology: A New Focus*. Vol. V, *Exploiters and the Exploited*. Lexington: Lexington Books.
Eidgenössische Materialprüfungs- und Forschungsanstalt (EMPA) (2009). 'Ewasteguide.info: Hazardous Substances in e-Waste.' Retrieved 23 August 2011, from file:///D:/Mypercent20Documents/PhD_transnatenvironmcrime-andgovernance/Hazardouspercent20Substancespercent20inpercent20e-Wastep ercent20percent20percent20ewasteguide.info.htm.
Elias, R. (1986) *The Politics of Victimization*. Oxford: Oxford University Press.
Environmental Investigation Agency (2011) *System Failure. The UK's Harmful Trade in Electronic Waste*. London: Environmental Investigation Agency.
European Commission (2011) *Recast of the WEEE Directive*. Retrieved 12 December 2011, from http://ec.europa.eu/environment/waste/weee/index_en.htm.
European Environment Agency (2009) *Waste without Border in the EU? Transboundary Shipments of Waste*. Copenhagen: European Environment Agency.
Fattah, E. (2010) 'The evolution of a young, promising discipline: sixty years of victimology, a retrospective and prospective look', S. G. Shoham, P. Knepper

and M. Kett (eds) *International Handbook of Victimology*. Boca Raton, FL: CRC Press, pp. 43–94.

Fischer, C., Hedal, N., Carlsen, R., Doujak, K., Legg, D., Oliva, J., Lüdeking Sparvath, S., Viisimaa, M., Weissenbach, T. and Werge, M. (2008) *Transboundary Shipments of Waste in the EU. Developments 1995–2005 and Possible Drivers*. Copenhagen: European Environment Agency – European Topic Centre on Resource and Waste Management.

Friedrichs, D. (2002) 'Occupational crime, occupational deviance, and workplace crime: Sorting out the difference', *Criminal Justice*, 2(3): 243–256.

Gibbs, C., McGarrell, E. F. and Axelrod, M. (2010) 'Transnational white-collar crime and risk. Lessons from the global trade in electronic waste', *Criminology and Public Policy*, 9(3): 543–560.

Gibbs, C. and Simpson, S. S. (2009) 'Measuring corporate environmental crime rates: progress and problems', *Crime, Law and Social Change*, 51(1): 87–107.

Gilbert, M. J. and Russell, S. (2002) 'Globalization of criminal justice in the corporate context', *Crime, Law and Social Change*, 38(3): 211–238.

Goodey, J. (2005) *Victims and Victimology: Research, Policy and Practice*. Harrow, England: Pearson Longman.

Greenpeace (2008a) *Chemical Contamination at e-waste Recycling and Disposal Sites in Accra and Korforida, Ghana*. Greenpeace, UK: Greenpeace Research Laboratories.

Greenpeace (2008b) *Poisening the Poor. Electronic Waste in Ghana*. Amsterdam: Greenpeace International.

Gunningham, N., Kagan, R. and Thornton, D. (2003) *Shades of Green: Business, Regulation and Environment*. Standford: Stanford University Press.

Hall, M. (2011) 'Environmental victims: challenges for criminology and victimology in the 21st century', *Varstvoslovje/Journal of Criminal Justice and Security*, 13(4): 371–391.

Heiss, R., Ruessink, H., Isarin, N., Koparova, M. and Grabiel, D. (2011) International hazardous waste inspection project at seaports: results and recommendations. *INECE 9th International Conference*. British Columbia: Canada.

Hillyard, P., Pantazis, C., Tombs, S. and Gordon, D. (eds) (2004) *Beyond Criminology: Taking Harm Seriously*. Manitoba: Fernwood Publishing.

Holley, C., Gunningham, N. and Shearing, C. (2012) *The New Environmental Governance*. Abingdon, New York: Routledge.

Huisman, W. (2009) 'Schending van internationaal recht door ondernemingen', in G. Vande Walle and P. Van Calster (eds), *De criminologische kant van ondernemen*. Den Haag, Boom Juridische Uitgevers, pp. 171–186.

IMPEL-TFS (2005) *IMPEL-TSF Threat Assessment Project, the Illegal Shipments of Waste among IMPEL Member States*. London: Jill Dando Institute, University College London.

Interpol (2009) *Electronic Waste and Organized Crime. Assessing the Links. Phase II Report for the Interpol Pollution Crime Working Group*. Interpol. Available at http://www.interpol.int/%E6%88%BF%E6%80%95en/Crime-areas/Environmental-crime/Resources

King, G., Keohane, R. and Verba, S. (1994) *Designing Social Inquiry: Scientific Inference in Qualitative Research*. Princeton, NJ: Princeton University Press.

Leefmilieu, natuur en energie (LNE) [Environment, Nature and Energy] (2010) *Afvalstromen in Vlaanderen: risicoprofiel*. [Waste flows in Flanders: Risk Profiles]. Brussels: LNE.

Long, M. A., Stretesky, P. B., Lynch, M. J. and Fenwick, E. (2012) 'Crime in the coal industry: implications for green criminology and treadmill of production', *Organization and Environment,* first published on 1 July, doi:10.1177/1086026612452266.

McEvoy, K. and McConnachie, K. (2012) 'Victimology in transitional justice: victimhood, innocence and hierarchy', *European Journal of Criminology,* 9(5): 527–538.

Nordbrand, S. (2009) *Out of Control. E-waste Trade Flows from the EU to Developing Countries.* Stockholm: SwedWatch.

Passas, N. (1999) 'Globalization, criminogenic asymmetries and economic crime', *European Journal of Law Reform,* 1(4): 399–423.

Passas, N. and Goodwin, N. (eds) (2004) *It's Legal but It Ain't Right: Harmful Social Consequences of Legal Industries.* Ann Arbor: University of Michigan Press.

Pearce, F. and Tombs, S. (1998) *Toxic Capitalism: Corporate Crime and the Chemical Industry.* Aldershot: Dartmouth Publishing Company.

Pellow, D. (2007) *Resisting Global Toxics: Transnational Movements for Environmental Justice.* Cambridge, MA: MIT Press.

Prakash, S. and Manhart, A. (2010) *Socio-economic Assessment and Feasibility Study on Sustainable E-waste Management in Ghana.* Freiburg: Öko-Instutut e.V.; Inspectorate of the Ministry of Housing, Spatial Planning and the Environment of the Netherlands (VROM-Inspectorate) and the Dutch Association for the Disposal of Metal and Electrical Products (NVMP).

Quadri, S. (2010) 'An analysis of the effects and reasons for hazardous waste importation in India and its implementation of the Basel Convention', *Florida Journal of International Law,* 22(3): 467.

Robinson, B. (2009) 'E-waste: an assessment of global production and environmental impacts', *Science of the Total Environment,* 408: 183–191.

Schluep, M., Rochat, D., Wanjira, A. M., Laissaoui, S. E., Wone, S., Kane, C. and Hieronymi, K. (2008) Assessing the e-waste situation in Africa. *Electronics Goes Green 2008+* (8–10 September Berlin). Available at http://www.ewasteguide.info

Schwarzer, S., De Bono, A., Giuliani, G., Kluser, S. and Peduzzi, P. (2005) *E-waste, the Hidden Side of IT Equipment's Manufacturing and Use.* United Nations Environment Programme. Retrieved from http://archive-ouverte.unige.ch/unige:23132

Sepúlveda, A., Schluep, M., Renaud, F. G., Streicher, M., Kuehr, R., Hagelüken, C. and Gerecke, A. C. (2010) 'A review of the environmental fate and effects of hazardous substances released from electrical and electronic equipments during recycling: examples from China and India', *Environmental Impact Assessment Review,* 30(1): 28–41.

Skinnider, E. (2011) *Victims of Environmental Crime – Mapping the Issues.* Vancouver: The International Centre for Criminal Law Reform and Criminal Justice Policy.

Slapper, G. and Tombs, S. (1999) *Corporate Crime.* Essex: Pearson Education Limited.

South, N. (1998) 'Corporate and state crimes against the environment: foundations for a green perspective in European Criminology', in V. Ruggiero, N. South and I. Taylor (eds) *The New European Criminology: Crime and Social Order in Europe.* London and New York: Routledge.

South, N. (2007) 'The corporate colonisation of nature: bio-prospecting, bio-piracy and the development of green criminology', in P. Beirne and N. South (eds)

Issues in Green Criminology. Confronting Harms against Environments, Humanity and Other Animals. Portland: Willan Publishing, pp. 230–247.
Stephens, S. (2009) 'Reflections on environmental justice: children as victims and actors', *Social Justice*, 23(4): 62–86.
Stover, E. and Weinstein, H. (eds) (2004) *My Neighbor, My Enemy: Justice and Community in the Aftermath of Mass Atrocity*. Cambridge: Cambridge University Press.
Strang, H. and Sherman, L. (2003) 'Repairing the harm: victims and restorative justice', *Utah Law Review*, 1: 15–42.
Stretesky, P. and Lynch, M. (1998) 'Corporate environmental violence and racism', *Crime, Law and Social Change*, 30(2): 163–184.
Stretesky, P. and Lynch, M. (2009) 'Does self-policing reduce chemical emissions?' *The Social Science Journal*, 46(3): 459–473.
United Nations Environment Programme (2005) *E-waste, the Hidden Side of IT Equipment's Manufacturing and Use*, see Schwarzer et al. (2005).
United Nations Environment Programme (2006) *Call for Global Action on E-waste*, United Nations Environment Programme available at: http://www.unep.org/documents.multilingual/default.asp?articleid=5447&documentid=496&l=en
van Erp, J. and Huisman, W. (2010) 'Smart regulation and enforcement of illegal disposal of electronic waste', *Criminology and Public Policy*, 9(3): 579–590.
Vande Walle, G. (2012) 'Conflict resolution or risk management?' in P. Ponsaers (ed.) *Social Analysis of Security – Financial, Economic and Ecological Crime – Crime, (In)security and (Dis)trust – Public and Private Policing*. The Hague: Eleven International Publishers, pp. 31–55.
von Hentig, H. (1948) *The Criminal and His Victim*. New Haven: Yale University Press.
VROM-inspectie (2011). *Evaluatie en vooruitblik ketenproject elektrische en elektronische apparaten 2010* [Evaluation and preview supply chain project electrical and electronic equipment 2010]. Den Haag, Volkshuisvesting, Ruimtelijke Ordening en Milieubeheer (VROM) Inspectie – Directie Uitvoering.
Walters, R. (2007) 'Crime, regulation and radioactive waste in the United Kingdom', in P. Beirne and N. South (eds) *Issues in Green Criminology. Confronting Harms against Environments, Humanity and Other Animals*. Portland: Willan Publishing, pp. 186–205.
White, R. (2007) 'Green criminology and the pursuit of social and ecological justice', in P. Beirne and N. South (eds) *Issues in Green Criminology. Confronting Harms against Environments, Humanity and Other Animals*. Portland: Willan Publishing, pp. 32–54.
White, R. (ed.) (2009) *Environmental Crime: A Reader*. Collumpton: Willan Publishing.
White, R. (2011) *Transnational Environmental Crime. Towards an Eco-global Criminology*. London and New York: Routledge.
Williams, C. (1996) 'An environmental victimology', *Social Justice*, 23(4): 16–40.
Yin, R. (2009) *Case Study Research: Design and Methods*. Thousand Oaks, CA: Sage.

Part II

Rights and Wrongs

3
Resource Wealth, Power, Crime, and Conflict

Avi Brisman and Nigel South

Introduction

The global flow of capital and competitive trading across borders has been accompanied by the weakening of the ability of regulators and sovereign countries to monitor and restrict harmful activities of multinational corporations. Multinationals often exert a disproportionately large amount of influence over the regulatory agencies that are charged with regulating them – a condition referred to as 'regulatory capture' (see Stigler, 1971 in Borak, 2011) – and they have increasingly taken advantage of these globalising circumstances to lower environmental standards and to collude in violation of the rights of inhabitants of threatened locations (human and nonhuman) and of activists seeking to protect the environment (see, for example, Boelens et al., 2011; Clark, 2009; Global Witness, 2012; Newell, 2001; Williams, 1996).

In countries such as Angola, the Democratic Republic of Congo (DRC), Liberia, Nigeria, Sierra Leone, and Zimbabwe, rather than deriving broad benefit from resource wealth, local populations have instead suffered from what has been called the 'resource curse' (see, for example, Auty, 1993; Auty and Mikesell, 1998; de Soysa, 2000; Ross, 2003a, b; Wenar, 2008). This has manifested itself as the damaging and divisive exploitation of environmental wealth in forms such as illegal trades in diamonds, timber, and wildlife that, in turn, have generated funds that have supported and perpetuated internal conflicts, corruption, and the externalising of economic surplus.[1] Curtailing both politically powerful elites' and rebel groups' access to revenues from high-value natural resources is vital to ensuring long-term peace as well as preventing catastrophic environmental degradation, but it is but one of the many ways in which conflict and the environment are linked.

Elsewhere, we have explored notions and sources of 'environmental rights' and have suggested that regardless of whether 'environmental rights' are conceptualised as distinct rights or as part of or an extension of the broader realm of human rights, environmental degradation and environmental protection are intertwined with both human rights and environmental rights (Brisman, forthcoming; South and Brisman, 2013). In this chapter, we begin with the position that the environment and natural resources can be *a source of conflict* (for example when groups fights over access to or use of natural resources), can *fuel or fund existing conflicts* (for example, when warring groups extract diamonds or metals or timber that are then sold to finance conflicts), and can be a *casualty of conflict* (for example, in the Vietnam War, when deforestation chemicals, such as Agent Orange, caused crop destruction; in the first Gulf War, when oil wells were set ablaze); see, for example, Brisman, 2008; Muffett and Bruch, 2011; South, 1998; Williams, 1996.

We seek to further demonstrate how environmental degradation frustrates the pursuit of social justice and the realisation of human rights (broadly conceived), and how the abuse of power and human rights abuses can lead to the exploitation of natural resources and environmental harm. We provide some illustrations of such scenarios in the context of discussion of power, politics, and resource security, before briefly considering how we might offer some protection for people and the planet in the face of the abuse of power.

Natural resources, crime, and conflict

Describing forms of 'environmental crime', Shover and Routhe (2005: 324) observe that 'it is not hyperbole to suggest that environmental crime can victimize entire populations or nations; to the extent that natural resources belong to a nation's people, theft or destruction of them victimizes all.' Recognising the far-reaching and cross-border nature of such victimisation and the operations of offenders, Shover and Routhe (2005) argue that, on a global basis, nations have begun to regulate and prohibit various activities and behaviours such as trades in rare species and ozone-depleting substances, overfishing and whaling, and hazardous waste transport and dumping.

At the same time, however, they recognise that this is not sufficient and that there are in any case limits to the efficacy of such initiatives and interventions. These limitations may be rooted in both public apathy and private interest. In the West, the general public may be ignorant of or uninterested in issues arising from environmental harms and breaches of human rights, and if they do become aware, may feel 'compassion fatigue' about yet another cause they are being asked to embrace (Cohen and Seu, 2002) or consumer hostility if remedy or rectification of wrongs may mean consumer items they favour are less accessible or more expensive. Within the countries affected, complexities related to group and local alliances, differentials in distribution of income and rewards, the paramount needs of survival and subsistence, and many other factors may mean there is a mix of disquiet, opposition, acceptance, or strong support regarding these activities.

Importantly, straddling the consumer nations and producer points of origin of resources are the multinational corporations – the parent companies and their subsidiaries, structured and operating in particular ways that insulate the parent from accusations of wrongdoing but keep profits within the corporate family (Mathiason, 2011). Such arrangements and interests generate powerful lobbying forces in both developed and developing nations and can be highly influential in shaping laws and rules that are in principle designed to regulate and control their conduct (Shover and Routhe, 2005: 346, 357; Yeager, 1987). Crucially, when trade agreements, regulatory protocols, and policies setting out rights and obligations come to the point of adoption and implementation, follow-through and adherence can be weak and variable (see Burnley, 2011). As Shover and Routhe (2005: 357) explain, this can occur for a number of reasons:

> Third world countries may be incapable of resisting or effectively controlling industries' self-interested interpretation of rules and their environmentally destructive actions. Corporate owners and managers can threaten to relocate to jurisdictions with less restrictive regulatory approaches, with the resulting loss of jobs or tax revenues. Police and prosecutors in most local jurisdictions and many nations do not have the budget, expertise or other resources needed to pursue cases of environmental crime aggressively. Issues of national sovereignty also complicate development of a coherent international environmental law and credible enforcement.

It is clear, then, that while there may seem to be considerable legal and political agreement within and between nations that cover and are embraced by corporate bodies regarding conduct that respects the environment as well as human rights, the temptations of profit and power that the gift of resource riches can produce may be too great. The tragic irony is that such riches do not always – or indeed, usually – bestow on either the general population or the macro-economies of these nations much in the way of wealth and stability (Ross, 2003a, b, 2004). As Ross (2003b: 5) observes, 'It may seem paradoxical that a "gift" from nature of abundant oil, gold or gemstones tends to cause economic distress. Yet study after study has found that resource-dependent economies grow more slowly than resource-poor economies'.

There are several common characteristics of states in such circumstances. For example, poverty rates are frequently high, and this is related to the tendency for resource-rich governments to 'do an unusually poor job of providing education and health care for their citizens' (Ross, 2003b: 6; see also Lujala and Rustad, 2011); for Angola, see, for example, Associated Press, 2012; for the Democratic Republic of Congo (DRC), see, for example, Burnley, 2011; for Liberia, see, for example, BBC, 2012a; Ford, 2012; for Namibia, see, for example, Fuller, 2011; for Nigeria, see, for example, Duffield, 2010. This can increase prospects of instability, insecurity, conflict, associated crime, and civil war. As Ross (2003b: 6–7) observes,

> Not surprisingly, people are more likely to rise up against their government when their economic predicament is bad and getting worse. Rebel groups find it easier to recruit new members when there is widespread poverty and unemployment, since it makes the prospect of combat and looting seem more attractive by comparison.

If wealth is managed efficiently and effectively, through processes of sound governance that are seen as legitimate and fair, there is a basis for stable political and socioeconomic life (see, for example, Lujala and Rustad, 2011; Muffett and Bruch, 2011). A high degree of natural resource dependence that produces high returns of wealth, however, can also have a distorting effect on governments and can lead to actions that are both criminal in themselves as well as criminogenic and can produce conditions favourable to conflict and civil war.

Ross (2003b: 8) identifies three mechanisms that may be at work here: corruption, state weakness, and reduced accountability. First, Ross cites strong evidence of correlations between high resource dependence and measures of corruption: 'Resource wealth often floods governments with more revenue than they can effectively manage' and this 'tends to be collected by governments in ways that are unusually difficult for citizens to track – and unusually easy for crooked officials to divert; hence some of it winds up in off-budget accounts or the pockets of government officials....'

Second, resource abundance provides a flow of finance as a 'gift' rather than as something that has to be created by other economic activity that, in turn, would be taxed to provide revenue to fund government and services. A government that does not have the elaborated bureaucracy required to collect taxation, which depends on a means of being in touch with the population, may therefore be weak in its ability provide public goods, or to effectively and consensually police society and reduce conflict. A third effect can follow from the second: Because acquiescence in a system of taxation and service provision is not needed where revenues flow easily from sales of natural resources, less attention may be paid to systems of democracy and voting, and the wealth that government accumulates can be used to 'buy off' the population by reducing or curtailing taxation and/or investing in army and security services that suppress political opposition and dissent on the streets (Ross, 2003b: 10–15; see also Associated Press, 2012).[2]

Various cases can be cited to illustrate these conditions and effects. For example, in 2001 alone, almost $1 billion was corruptly diverted from the accounts of the Angolan government according to the International Monetary Fund (Ross, 2003b: 9–10). Sierra Leone provides an illustration of how mineral wealth spread across a large geographical territory can pose great challenges to governments in terms of territorial control, with failings in the licensing of mining and the emergence of protection services for miners provided by armed gangs and private armies, eroding legitimacy and creating abrasive relations; civil war followed (see generally Le Billon, 2011). In the case of the DRC, rich in mineral resources such as coltan (columbite-tantalite), gold, tungsten, and tin ore used for jewellery, mobile phones, and laptops, the country has suffered ceaseless conflict for nearly two decades, has seen ongoing factional warfare among and between the Congolese army, 'defence forces', and 'rebel units', and has felt the effects of genocide in neighbouring Rwanda (Adams, 2011; see also Burnley, 2011; Draper, 2011; Le Billon, 2011). According to a UN expert panel, this led to 'highly organized and systematic exploitation' (United Nations, 2002: 10, 52) involving, *inter alia*, rebels and government forces profiting from the trades in mineral ores; subjecting civilians to massacres, rape and extortion; and using forced labour and coercing children into the role of soldiers. In addition, although it was traditionally the case that women were engaged in farming, the fields are in forests that are now occupied by rebels and growing food has become too dangerous. War has brought a particular kind of devastation: deaths arising from such conflicts have not been limited to combatants, but have included high proportions of civilians.

Perhaps 40 per cent of 'war casualties' in the DRC at some points have been women and children, as hospital and health care services collapse and preventable disease and malnutrition increase mortality (Montague, 2002). Life expectancy and environmental conditions are also eroded by practices such as child labour, with recent evidence of children as young as 10 working 150 feet underground in the Tilwezembe copper mine in the DRC, and acidic waste resulting from copper refining at Luilu in Katanga Province being pumped directly into the Luilu River (Sweeney, 2012: 16).[3]

To add another level of complexity, Kuijpers (2012) points out that there is also a classic criminal motivation at work within the set of drivers of conflict and war – greed. As Kuijpers (2012: 14) explains,

> Although it is often assumed that conflicts occur because of grievance, driven by high inequality, a lack of political rights, or ethnic and religious divisions,... many conflicts can better be explained by economic variables and... greed is a better explanatory factor for conflict than grievance.

Although Lujala and Rustad (2011) assert that 'resource capture' may be a goal of rebel uprisings or violent secessionist movements, personal or regional greed and enrichment are rarely if ever the sole motivation for predatory, greed-driven armed rebellion. Le Billon (2011: 16), on the other hand, maintains that some combatants are, indeed, 'drawn to rebellion by short-term, opportunistic economic objectives rather than by long-term political objectives'. Although the extent to which greed is a motivating factor – or *the* motivating factor – may be subject to debate, examples certainly exist.

For instance, in oil-rich Nigeria, where enforcement of environmental regulations is lax (in part because multinational oil companies such as Shell and ExxonMobil work in tandem with the state oil firm), armed gangs frequently damage pipelines to steal crude (BBC, 2012b; Duffield, 2010; Hirsch and Vidal, 2012; Owolabi, 2012; Ross, 2012), leading to allegations of both damage to the environment and human rights abuses. In the case of the damage to the environment, sabotage and militant attacks, as well as illegal refining and oil theft by impoverished Nigerians seeking more of the benefits of Nigeria's natural resources, have resulted in thousands of barrels of oil spilled each year, on top of those caused by the oil companies' own activities, which, together, have transformed parts of the Niger Delta – one of the largest river deltas in the world – into an oily wasteland replete with dead mangroves and fish (BBC, 2012b; Owolabi, 2012; Ross, 2012). With regard to the erosion of human rights, Shell, in an effort to protect its operations, has spent hundreds of millions of dollars over a three-year-period – more than a third of its security budget, by some accounts – on private security forces and supplies (for example, gunboats, helicopters, and other vehicles and satellite phones) to Nigerian police and government forces known for routine human rights abuses (Hirsch and Vidal, 2012; see also Ross, 2012; Williams, 1996). In the DRC, to offer another example, control of mines generating huge revenues is a substantial financial motive for aggressive actions.

Of course, it should not be overlooked that there are some benefits of resource wealth for the ordinary populations of the DRC and similar nations – for example, to provide secure income when agriculture is no longer pursued or presents risks (Kuijpers, 2012: 17; Perks and Vlassenroot, 2010: 71; see also BBC, 2012a; Ford, 2012; Lujala and Rustad, 2011). But while local populations may live on the land that provides these gifts from nature and the planet, these people are not its real beneficiaries. Thus, we might conclude that natural resource wealth can create criminality, stimulate civil war and provide incentives for external economic agents to enter and exploit the resource market.[4]

Power, politics, and resource security

In the rich West, whether it is in our everyday consumption of meat and soy from Brazil (see, for example, Ettinger, 2011; Nepstad, Stickler and Almeida, 2006) or our habitual use of smartphones, laptops, and digital cameras (or even 'green machines', such as hybrid cars and wind turbines) containing metals from China and the DRC (see, for example, Folger, 2011; Krugman, 2010; Risen, 2010; Weatherford, 2011), individually and collectively we are 'a link in a long global chain' (Polgreen, 2008: 1) – one in which, as is especially the case with the West's relationship to Africa, 'the richest countries...often treat the poorest less as partners in progress than as cheap targets for resource extraction' (Owen, 2012: 74). As prolific consumers, we are contributors to environmental degradation and human rights abuses in these regions, committing our 'ordinary acts' of everyday ecocide (Agnew, 2013). This is a 'global chain' on which Western economies are dependent and these patterns of consumption are accepted as essential to support 'business as usual' (Fussey and South, 2012) and a standard of living that must be guaranteed. This can be seen, for example, in the political discourses of the Bush administration which, as Lynch and colleagues (2010) demonstrate, aimed to undermine climate change evidence and reassure consumers about the acceptability of carbon-profligate lifestyles. For the Bush administration and corporate collaborators, this was about defending economic interests and 'the preservation of a way of life' (see also Brisman, 2005, 2012). In this way, the preservation of the supply chain and the natural resources that feed it become not matters of international environmental sustainability but of 'national security' and 'national interest'.

In the post–Cold War era of globalisation, national security strategists have seized on the significance of 'environmental security' as a matter that should be concerned with responses to 'population pressures, resource scarcities and intrastate conflict in the South' (Hartmann, 1999) and containment and intervention initiatives that serve the national (Northern/Western) interest (South, 2012). Keenan (2009), for example, shows how flimsy evidence of terrorist activity linked to kidnappings in Algeria in 2003 provided the Bush administration with its justification to increase a military presence in Africa, with the real aim being to secure access to resource wealth, in particular African oil in countries like Algeria, Nigeria, and Libya. Perhaps the 'resource curse' is beginning to extend its effects in even more dangerous ways making restraints on the limits and abuse of power even more urgently needed than ever.

Protection for people and the planet in the face of the abuse of power: a few considerations

As outlined above, conflict can stem from disagreements over access to and the use, distribution, and governance of finite natural resources. Natural resources can also provide financing for groups seeking to start or resume armed conflict. Thus, according to Lujala and Rustad (2011: 20), 'High-value natural resources have been associated with dozens of armed conflicts, millions of deaths, and the collapse of several peace processes.' At the same time, the environment is frequently a casualty of conflict, where violent contestations have scorched the earth, eradicated species, destroyed entire ecosystems, and rendered large swathes of land uninhabitable. Neighbouring regions, which do not possess the sought-after natural resource and are not the locus of fighting, can also be affected. As Muffett and Bruch (2011: 4) explain, 'People displaced by conflict can be drawn together into informal tent cities or organised encampments numbering in the hundreds of thousands. These settlements can become major urban areas virtually overnight, requiring a steady supply of fresh water, sanitation facilities, fuel wood, building supplies and food that far exceeds local resources.'

While the linkages between conflict and the environment are varied, they are not all negative – or, to put it another way, they do not always describe or represent current or anticipated violence. Muffett and Bruch (2011: 4) contend that 'well-managed resources can help fund reconstruction efforts and help bring order from chaos.' Similarly, Lujala and Rustad (2011: 19) argue that

> High-value natural resources have the potential to promote and consolidate peace.... Valuable resources can help to jump-start development, secure sustainable growth, raise living standards, and increase economic equality. They are also an important source of foreign currency for cash-strapped governments, can reduce dependence on international aid, and can support compensation and post-conflict relief from war-affected populations.

Outlining the foundations of a theory of intergenerational ecological justice, Weston (2012: 261) draws on Whitehead in emphasising that the community of humankind is made up of generations of the present, the past, and the future, with the related implication that rights and obligations should also apply across this long intergenerational chain: 'In this manner, the "common heritage" of Earth's natural resources, fresh water systems, oceans, atmosphere, and outer space belongs to all generations in an inter-temporal partnership.' Systems and statements of law, governance and rights should build on respect for the interdependence of ecosystems and the principle of intergenerational equity. How then might we seek to guarantee the protection of a common heritage and undo the exploitation and injustices that have accompanied resource wealth? How do we ensure that high-value natural resources serve as assets to whole populations here and now and in the future, rather than just conferring benefits in the present to small groups? How can adverse environmental effects of resource extraction be minimised without damaging livelihoods, thwarting long-term development objectives, and impeding macroeconomic growth (or recovery)?

Whether on national or international levels, prevailing modes of regulation and governance tend to be weak, frequently shaped by ideologies of deregulation and a belief in voluntary compliance. Legislatively empowered but toothless regulators oversee processes that, via legal debate and mitigation, end up with the trivialisation of offences. Huisman (2010: 56) points out that these 'soft law' instruments can 'contribute to creating generally accepted social norms' underpinning protection of human rights or the environment by corporations, but that 'the worst offenders are not compelled to take part' while 'increasing numbers of corporations affiliated to the UN Global Compact initiative do not comply with their reporting obligations.'

Stewart (2011) argues that the 'resource curse' affecting nations that are rich in terms of resource endowment, but poor in terms of social development and most prone to violent upheaval can be addressed by curtailing corruption, regulating resource industries domestically, and enhancing judicial capacity in countries recovering from war. But the liability of foreign businesses for trading in illicit commodities (whether conflict minerals or endangered wildlife) must also be recognised, and this requires an effective legal tool. Here the concept of 'pillage' has been advanced by Stewart (2011: 11–14) and the Open Society Justice Initiative as the basis for the more rigorous prosecution of offences that can be identified as part of this post–Cold War plunder of minerals, metals, timber, and other natural resources.[5] 'Pillage' has a long history of application in describing offences under the laws of war and as an element of various war crime statutes.[6] Because it has been used against former politicians and in post–World War II prosecutions of Nazi business leaders for theft of property from occupied countries, but has rarely been used against modern corporations or their officials, the proposal is that on the basis of similarities between these earlier crimes and corporate practices in the resource wars affecting countries today, it is time to put these ideas to the test.

One further important perspective and tool already applied in various cases of contestation and conflict is the idea of restorative justice. This is now seen to hold considerable promise as a means to resolve responsibility and compensate for abuse of rights and crimes against the environment and the human and nonhuman beings affected. The idea of restorative justice is, in one sense, quite simple and has a long history, but as Braithwaite (2002: 7–8) notes, the important point for his purposes – and indeed for ours – is that we have witnessed 'a late modern revival of restorative justice that has its deepest roots in a shift from most regulatory activities, having individuals and their bodies as their objects to a world where more of the wrongdoing is done by organisations that are regulated in a mostly restorative fashion.'

Although we cannot develop the point here, this kind of approach to the administration of environmental justice, invoking methods and principles of mutual engagement and shared learning, is both practical as well as consonant with green ideals.

Conclusion

Ultimately, we must recognise that because natural resources have different characteristics (for example, some are more easily extracted, concealed, smuggled, and sold than others; some are scarce, existing in only a small number of countries and have few, if any, substitutes; some exist within nation-states, while others cross or are located at nation-state boundaries), the degree, extent, and ways in which they are relevant to conflict may vary (see Lujala and Rustad, 2011). As such, there is no one size fits all approach to preventing conflict or ensuring postconflict peace in resource-rich countries. Attention to context – what Muffett and Bruch (2011: 6) call 'situational awareness' – is crucial. That said, one constant appears to be that proper management of high-value natural resources (and their associated revenues) – which involves increased and/or improved 'democratic inclusion' (Muffett and Bruch, 2011: 6) – is vital to averting or ending natural resource-related conflict. But here, too, the actual form of 'democratic inclusion' must also be developed, understood, and analysed on a case-by-case basis, and we would not begin to suggest that the kind of successes achieved through 'collaborative' land and natural resource management in the western United States (see Kemmis and McKinney, 2011), for example, can or should be replicated as part of water management reform in Central Asia (see Bichsel, 2011) or Latin America (see Boelens et al., 2011). Nor do we see a piecemeal approach to improved governance on local, national, and international levels as a panacea. The inclusion of workers, the poor, and indigenous peoples in the management of high-value natural resources may forestall or mitigate environment-related conflict. In the present state of global economic failings, however, no amount of democratic development in these resource-producer nations can overcome or compensate for Western complicity in the unsustainable production practices and exploitative capitalist consumerism that degrade the environment and destabilise resource-rich regions.

Notes

1. Externalisation of profits is a widespread problem in Africa. For example, Zimbabwe provides a case of the failure of a regulatory system – the Kimberley Process – that was designed to prevent profiteering from conflict diamonds and prevent the use of such profits to fund the violence of President Mugabe's regime that is directed against political opponents (McVeigh, 2011: 22). For recent debate in Zambia, see Lusaka Times/Times of Zambia (2012).

2. Along these lines, Lujala and Rustad (2011: 20) explain that '[a] government that is able to finance its budget through natural resource revenues rather than public taxation can easily become detached from, and therefore less accountable to, the populace'.
3. According to Le Billon (2011: 16), rebel groups operating in resource-rich environments tend to commit particularly awful abuses against civilians.
4. For an argument that the conflict over the allocation of natural resources in the Permian Basin area in southeast New Mexico and west Texas is driven by 'economic greed and political power', see Peterson et al. (2011: 35).
5. Le Billon (2011: 14) explains that 'the importance of resources to armed groups has grown rapidly since the late 1980s [and the end of the Cold War], as belligerents turned to natural resources to replace external political sponsorship'. Similarly, Lujala and Rustad (2011: 21) observe that 'since the end of the Cold War, financing from the superpowers has declined and revenues from valuable natural resources has gained importance as a source of conflict financing'.
6. According to Stewart (2011: 15), although the terms 'pillage', 'plunder', 'spoliation', and 'looting' are all commonly used in legal discussion with more or less the same meaning, 'pillage' is the only one that features in treaties governing the laws of war.

References

Adams, T. (2011) 'How the teachers of hope I met in the Congo were brutally killed', *The Observer*, 16 October. Accessed at: http://www.guardian.co.uk/world/2011/oct/16/teachers-of-hope-in-congo-brutally-killed.

Agnew, R. (2013) 'The ordinary acts that contribute to ecocide: a criminological analysis', in N. South and A. Brisman (eds) *The Routledge International Handbook of Green Criminology*. London: Routledge.

Associated Press (2012) 'Thousands stay away from Angolan elections', *National Public Radio (NPR) Broadcast*, 31 August. Accessed at: http://www.npr.org/templates/story/story.php?storyId=160366826.

Auty, R. M. (1993) *Sustaining Development in Mineral Economies: The Resource Curse Thesis*. London: Routledge.

Auty, R. M. and Mikesell, R. F. (1998) *Sustainable Development in Mineral Economies*. Oxford: Oxford University Press.

BBC (2012a) 'Liberia' rainforest: '60 percent handed to logging companies', *BBC News Africa*, 4 September. Accessed at: http://www.bbc.co.uk/news/world-africa-19469571.

BBC (2012b) 'Shell urged to pay Nigeria $5bn over Bonga oil spill', *BBC News Africa*, 17 July. Accessed at: http://www.bbc.co.uk/news/world-africa-18875731.

Bichsel, C. (2011) 'Liquid challenges. Contested water in Central Asia', *Sustainable Development Law and Policy*, XII(1): 24–30.

Boelens, R., Bueno de Mesquita, M., Gaybor, A. and Peña, F. (2011) 'Threats to a sustainable future: water accumulation and conflict in Latin America', *Sustainable Development Law and Policy*, 12(1): 41–45, 67–69.

Borak, M. (2011) 'Weak planning process frustrates protection of Puerto Rico's threatened coastline', *Sustainable Development Law and Policy*, 12(1): 23, 51, 58.

Braithwaite, J. (2002) *Justice and Responsive Regulation*. New York: Oxford University Press.

Brisman, A. (2005) 'The aesthetics of wind energy systems', *New York University Environmental Law Journal*, 13(1): 1–133.

Brisman, A. (2008) 'Crime-environment relationships and environmental justice', *Seattle Journal for Social Justice*, 6(2): 727–817.

Brisman, A. (2012) 'The cultural silence of climate change contrarianism', in R. White (ed.) *Climate Change from a Criminological Perspective*. New York: Springer.

Brisman, A. (Forthcoming [summer 2013]) 'Environmental and human rights: land, property and crime', in G. Bruinsma and D. Weisburd (eds) *Encyclopedia of Criminology and Criminal Justice*. New York: Springer Verlag.

Burnley, C. (2011) 'Natural resources conflict in the Democratic Republic of the Congo: a question of governance', *Sustainable Development Law and Policy*, 12(1): 7–11, 52–53.

Clark, Richard D. (2009) 'Environmental disputes and human rights violations: a role for Criminologists', *Contemporary Justice Review*, 12: 129–146.

Cohen, S. and Seu, B. I. (2002) 'Knowing enough not to feel much: emotional thinking about human rights appeals', in M. Bradley and P. Pedro (eds) *Truth Claims: Representation and Human Rights*. Piscataway, NJ: Rutgers University Press.

de Soysa, I. (2000) 'The resource curse: are civil wars driven by rapacity or paucity?' In M. Berdal and D. M. Male (eds) *Greed an Grievance: Economic Agendas in Civil Wars*. Boulder, CO: Lynne Reinner.

Draper, R. (2011) 'Rift in paradise', *National Geographic*, November: 82–113.

Duffield, C. (2010) 'Who are Nigeria's mend oil militants?' *BBC News Africa*, 4 October. Accessed at: http://www.bbc.co.uk/news/world-africa-11467394.

Ettinger, J. (2011) 'Vegan mock meats: to eat or not to eat', *Organic Authority*. Accessed at: http://www.organicauthority.com/foodie-buzz/vegan-fake-meat-should-you-or-shouldnt-you.html#s.abowoczbaayaa. Accessed on 2 April 2013.

Folger, T. (2011) 'Rare earths: they're the secret (Chinese) ingredients of (almost) everything high-tech', *National Geographic*, June: 136–145.

Ford, T. (2012) 'Liberia's failed logging promises', *BBC News Africa*, 4 September. Accessed at: http://www.bbc.co.uk/news/world-africa-19469570.

Fuller, A. (2011) 'Africa's super park', *National Geographic*, June: 60–77.

Fussey, P. and South, N. (2012) 'Heading toward a new criminogenic climate: climate change, political economy and environmental security', in R. White (ed.) *Climate Change from a Criminological Perspective*. New York: Springer Verlag.

Global Witness (2012) *Signing Their Lives Away: Liberia's Private Use Permits and the Destruction of Community-Owned Rainforest*, September. Accessed at: http://www.globalwitness.org/sites/default/files/library/Signingpercent20theirpercent20Livespercent20awaypercent20-percent20Liberianpercent20Privatepercent20Usepercent20Permitspercent20-percent204percent20Septpercent202012_0.pdf. Accessed on 5 September 2012.

Hartmann, B. (1999) 'Population, environment and security: a new trinity', in J. Silliman and and Y. King (eds) *Dangerous Intersections: Feminism, Population and the Environment*. London: Zed Books.

Hirsch, A. and Vidal, J. (2012) 'Shell spending millions of dollars on security in Nigeria, leaked data shows', *The Guardian* (UK), 19 August. Accessed at: http://www.guardian.co.uk/business/2012/aug/19/shell-spending-security-nigeria-leak.

Huisman, W. (2010) *Business as Usual? Corporate Involvement in International Crimes*. The Hague, NL: Eleven International Publishing.

Keenan, J. (2009) *The Dark Sahara: America's War on Terror in Africa*. London: Pluto Press.

Kemmis, D. and McKinney, M. (2011) 'Collaboration and the ecology of democracy', *Sustainable Development Law and Policy*, 12(1): 46–50, 69–70.

Krugman, P. (2010) 'Rare and foolish', *The New York Times*, 18 October: A35.

Kuijpers, K. (2012) 'The ambiguous effects of coltan mining in the Kivus', in M. Ganzinga, R. de Haan and K. Kuijpers (eds) *The Commodity Chain of Coltan from the Democratic Republic of Congo*, unpublished Criminological Journal, Utrecht University.

Le Billon, P. (2011) 'Bankrupting peace spoilers: what role for UN peacekeepers?' *Sustainable Development Law and Policy*, 12(1): 13–17, 54–55.

Lujala, P. and Rustad, S. A. (2011) 'High-value natural resources: a blessing or a curse for peace?' *Sustainable Development Law and Policy*, 12(1): 19–22, 56–57.

Lusaka Times/Times of Zambia (2012) *Zambia to Introduce Policy Measure to Stop Foreign Companies Externalising All Their Earnings-Chikwanda*, 27 June. Accessed at: http://www.lusakatimes.com/2012/06/27/zambia-introduce-policy-measure- stop-foreign-companies-externalising-earningschikwanda/.

Lynch, M., Burns, R. and Stretesky, P. (2010) 'Global warming and state-corporate crime: the politicization of global warming under the Bush Administration', *Crime, Law and Social Change*, 54: 213–239.

Mathiason, N. (2011) 'Human rights abuses? Blame the parents', *The Guardian*, Poverty Matters Blog, 9 December. Accessed at: http://www.guardian.co.uk/global-development/poverty-matters/2011/dec/09/human-rights-abuses-blame-parents.

McVeigh, T. (2011) 'Blood diamond "betrayal" helps Mugabe fund new election terror', *The Observer*, 18 December: 22.

Montague, D. (2002) 'Stolen goods: coltan and conflict in the Democratic Republic of Congo', *SAIS Review*, 22(1): 103–118.

Muffett, C. and Bruch, C. (2011) 'Introductory comments: the pervasive, persistent, and profound links between conflict and the environment', *Sustainable Development Law and Policy*, 12(1): 4–6.

Nepstad, D. C., Stickler, C. M. and Almeida, O. T. (2006) 'Globalization of the Amazon soy and beef industries: opportunities for conservation', *Conservation Biology*, 20(6) [December]: 1595–1603.

Newell, P. (2001) 'Managing multinationals: the governance of investment for the environment', *Journal of International Development*, 13: 907.

Owen, D. (2012) 'Dept. of invention: the artificial leaf', *The New Yorker*, 14 May: 68, 70–74.

Owolabi, T. (2012) *Oil Spill Stretches for Miles by Exxon Nigeria field*, Reuters, 1 September. Accessed at: http://news.yahoo.com/oil-spill-stretches-miles-exxon-nigeria-field-081809554.html.

Perks, R. and Vlassenroot, K. (2010) 'From discourse to practice: a sharper perspective on the relationship between minerals and violence in DR Congo', in J. Cuvelier (ed.) *The Complexity of Resource Governance in a Context of State Fragility:*

The Case of Eastern DRC. London: International Alert. Accessed at: http://www.international-alert.org/resources/publications/complexity-resource-governance-context-state-fragility-case-eastern-drc.

Peterson, L., Lininger, J. C., Bergoffen, M., Snape, B. and Bradley, C. (2011) 'Natural resource "conflicts" in the U.S. southwest: a story of hype over substance', *Sustainable Development Law and Policy*, 12(1): 32–35, 61–63.

Polgreen, L. (2008) 'CoNGOs riches, looted by Renegade troops', *The New York Times*, 16 November: 1, 14–15.

Risen, J. (2010) 'World's mining companies covet Afghan riches', *The New York Times*, 18 June: A4.

Ross, M. (2003a) 'The natural resource curse: how wealth can make you poor', in I. Bannon and P. Collier (eds) *Natural Resources and Violent Conflict: Options and Actions*. Washington, DC: World Bank.

Ross, M. (2003b) *Natural Resources and Civil War: An Overview*, World Bank. Accessed at: http://www.unepfi.org/fileadmin/documents/conflict/ross_2003.pdf.

Ross, M. (2004) 'What do we know about natural resources and civil war?' *Journal of Peace Research*, 41(3) [May]: 337–356.

Ross, W. (2012) 'Nigeria's booming illegal oil refineries', *BBC News Africa*, 26 July. Accessed at: http://www.bbc.co.uk/news/world-africa-18973637.

Shover, N. and Routhe, N. (2005) 'Environmental crime', in M. Tonry (ed.) *Crime and Justice: A Review of Research*. Chicago: University of Chicago Press.

South, N. (1998) 'A green field for criminology? A proposal for a perspective', *Theoretical Criminology*, 2: 211–233.

South, N. (2012) 'Climate change, environmental (in)security, conflict and crime', in S. Farrall, D. French and T. Ahmed (eds) *Climate Change: Legal and Criminological Implications*. Oxford: Hart.

South, N. and Brisman, A. (2013) 'Critical green criminology, rights and crimes of exploitation', in S. Winlow and R. Atkinson (eds) *New Directions in Crime and Deviance*, pp. 99–110. London: Routledge.

Stewart, J. (2011) *Corporate War Crimes: Prosecuting Pillage of Natural Resources*. New York: Open Society Foundation.

Stigler, G. J. (1971) 'The theory of economic regulation', *The Bell Journal of Economics and Management Science*, 2(1): 3–21. Accessed at: http://web.missouri.edu/~podgurskym/Econ_4345/syl_articles/Stigler_TheTheoryOfEconomicRegulation.pdf.

Sweeney, J. (2012) 'Mining giant in child labour and acid dumping claims', *The Observer*, 15 April: 16.

United Nations (2002) *Interim Report of the Panel of Experts on the Illegal Exploitation of Natural Resources and Other Forms of Wealth of the Democratic Republic of the Congo*, United Nations Document S/2002/565/22, 1–19 May. Accessed at: http://www.unhcr.org/refworld/pdfid/3d0471ad4.pdf.

Weatherford, K. (2011) 'The real cost of China's rare Earth export quotas on American job security', *Sustainable Development Law and Policy*, 12(1): 18, 55.

Weston, Burns H. (2012) 'The theoretical foundations of intergenerational ecological justice: an overview', *Human Rights Quarterly*, 34(1): 251–66.

Wenar, L. (2008) 'Property rights and the resource curse', *Philosophy and Public Affairs*, 36(1): 2–32.

Williams, C. (1996) 'An environmental victimology', *Social Justice*, 23: 16–40.

Yeager, P. (1987) 'Structural bias in regulatory law enforcement: the case of the U.S. Environmental Protection Agency', *Social Problems*, 34: 330–344.

4

Animal Trafficking and Trade: Abuse and Species Injustice

Ragnhild Sollund

Introduction

The illegal wildlife[1] trade is estimated to be the second-largest illegal trade worldwide (Warchol, 2007; Zimmerman, 2003; South and Wyatt, 2011), and it is steadily increasing (for example Smith, 2010; Stoett, 2002; Traffic, 2008), due to a globalised and expanded market in which the World Wide Web plays a significant role as an intermediary between offers and demands (IFAW, 2008). The illegal wildlife trade threatens one third of the world's species (Rivalan et al., 2007); the best-known species are the rhinoceros (for its horn) and the elephant (for its tusks) (Wasser, Clark and Laurie, 2009). In this chapter, I will first give a brief overview of the phenomenon, with a special focus on the parrot and reptile trades. Then I will show how a Norwegian case study reflects international findings and how the trade in endangered species in this country may be related to the international market. As reptiles are forbidden in Norway, this provides an interesting case for discussing problems of legalisation and regulation of the trade through the Convention on International Trade in Endangered Species of Wild Flora and Fauna (CITES Convention) versus criminalisation. Finally, I will discuss the trade in nonhuman species from a harm and justice perspective.

Large numbers of nonhuman animals are trafficked yearly, many of whom are directed to the international pet markets, including to collectors who keep exotic animals in private zoos as status symbols (Sollund, 2011; Pires and Clark, 2011b; Herbig, 2009). The illegal wildlife trade is addressed by criminologists who have analysed this trade in relation to organised crime and have examined its consequences in terms of harm and species extinction as well as crime prevention, usually under the umbrella of green, ecoglobal criminology (Zimmerman, 2003; Warchol,

Zupan and Clarke, 2003; Lemieux and Clarke, 2009; Schneider, 2008; Wilson-Wilde, 2010; Wyatt, 2009, 2011; Wellsmith, 2010, 2011; South and Wyatt, 2011; Pires and Clarke, 2011a, b; Pires and Moreto, 2011; Wright et al., 2001; Wyatt, 2009, 2011). The *illegal* wildlife trade is estimated to be from US $6 billion a year (Warchol, 2007), to US $10 billion a year (Schmidt, 2004), and even as high as US $20 billion (Alacs and Georges, 2007; South and Wyatt, 2011), but the legal trade is estimated to be from US $5 billion to $50 billion a year (Reeve, 2002: 10) to US $159 to $160 billion a year (Warchol, 2007; Duffy, cited in White, 2011: 55; Schneider, 2008). Whether the trade is classified as illegal or not depends on the degree to which the species is threatened with extinction.

These highly uncertain and varying economical estimates say nothing though, about the harm, abuse, and death inflicted on innocent animals, and in my view it would be wrong to highlight only the financial gains or losses involved in the trade, as this may make one lose sight of and perpetuate the real problem at hand: the massive suffering, species extinction, loss of biodiversity, and destruction of entire ecosystems produced by the unjust and anthropocentric abuse of power by humans. As the financial gains in most cases are the motives for the trade, however, they are important to bear in mind.

International wildlife trade, whether in animals or plants, is regulated in the CITES Convention (Convention on International Trade in Endangered Species of Wild Flora and Fauna), now signed by 175 member countries.[2] Species are listed on three appendices, according to the degree of threat to the species' survival. Trade in individuals who belong to species that are listed on the appendices must either be banned, except under very special circumstances (appendix 1), or must be accompanied by import and export (re-export) permits (appendices 2, 3), and species will, after conferences, be moved between appendices as the surviving numbers decrease or increase. It is urgent to emphasise that the purpose of the Convention is *not* to prevent the trade and trafficking, but to *regulate* it, the goal being to secure *sustainable* trade and species survival: 'Its aim is to ensure that international trade in specimens [sic] of wild animals and plants does not threaten their survival.'

CITES can be criticised for legitimating trade and trafficking in animals and for prolonging and encouraging abuse and species decline by regarding nonhuman species as exploitable resources for human use (Sollund, 2011; Hutton and Dickson, 2000). It is thus necessary to bear in mind the rationale behind the practices that maintain and prolong the animal trade, which are those related to culture, speciesism, and anthropocentrism (Sollund, 2011), which imply ideologies and convictions that are justifications for humans' discriminatory exploitative use of nonhuman animals (Nibert, 2002).

The parrot and reptile markets and ethical implications

Animals are traded for a large number of reasons. The principal endpoint for reptiles and parrots, however, is usually private homes, where they are held captive as so-called pets. Although parrots are legal in Norway, reptiles are not. This invites discussion about the moral legitimacy of why some threatened species can legally be kept as pets, while others cannot be. Without doubt, there are many so-called pets suffering in captivity (see, for example, Agnew, 1998; Maher and Pierspoint, 2011; Flynn, 2011). This is also the case for parrots and other birds; if they are not suffering from direct abuse, they suffer from neglect and malnutrition. Most wild-caught parrots in Costa Rica die after only a few years in captivity (Herrera and Hennessey, 2007; see Beirne, 1999 for a discussion of various forms of animal abuse). Without doubt, many parrots suffer in small cages with owners who are ignorant of their needs and their intelligence, as demonstrated by psychologist Irene Pepperberg in the case of the cognitive skills of African Grey parrots (Pepperberg, 1999).

Still, if provided sufficient (although not desirable) living conditions, many parrots can, like humans, live very long lives, and they are demanding and intelligent beings. Consequently, they are often passed from hand to hand. In the United States, 3 per cent of households keep birds; in Norway, 8 per cent (Mejdell, 2004)[3] of households keep a caged bird, and these markets, like those in other countries, have contributed to the threat to many species (Gonzales, 2003; Weston and Memon, 2009; Pires and Clark, 2009, 2011a). Many parrots in Norway must have arrived before the Norwegian ban, started when Norway joined CITES in 1976, as some species – for example, the popular blue-fronted Amazon – may reach up to 80 years of age. It is impossible to say whether parrots are wild-caught, as no one is obliged to identify his parrots.

In Mexico, 20 of 22 species of parrot are threatened with extinction because of loss of habitat and abduction (poaching) (Pires and Clarke, 2011a, b). There are also huge markets for parrots locally, in the countries where the parrots originate. In Costa Rica, 20 per cent of the population keep parrots, mostly wild-caught birds, because they are cheaper than hand-bred ones (Herrera and Hennesey, 2007). The very high mortality rate for parrots and other birds who are captured is partly caused by the methods used; for example, up to 50 per cent die during capture when trees with nests are felled (Gonzales, 2003). A report about the parrot trade in Mexico revealed that 65,000 to 78,500

Mexican parrots are captured each year and that the mortality rate for captured parrots exceeds 75–90 per cent before they reach the buyer (Guzmán et al., 2007; Michaels, 2002); see Wyatt, 2009 and Yi-Ming et al., 2000, in the case of raptor birds). The so-called 'harvesting' in some places is so extensive that total cohorts of birds are taken from their nests in the breeding period as a cultural routine, called 'La loreada' (Gonzales, 2003; see also Herrera and Hennessy, 2007 and Weston and Memon, 2009). Nests are destroyed and they aren't available for future generations. There is little reason to believe that keeping parrots and other birds (and animals) in captivity is better or more ethical by any significant standards than keeping reptiles. Still, some reptile species that naturally have small habitats and move little may suffer less in captivity than birds do. The extent of their frustration and suffering in captivity varies with the species they belong to and the conditions under which they are kept.

Recent research (Leal and Powell, 2011) on anole lizards, which often are used as pets, documents that the cognitive skills and behavioural flexibility of these reptiles far exceeds what was previously believed about them. Anoles can reverse previously learned associations. This experiment shows that reptiles are not stupid, as reptile keepers seem to think. My interviewees (see next section) would typically talk about their reptiles as being 'not very intelligent'. This, however, rather serves as an example of the ways in which humans very often fail to acknowledge the abilities of animals because of prejudice, social distance, and ignorance (Sollund, 2008).

Despite the value of spreading knowledge about the cognitive skills of reptiles, the above anole lizard experiment, together with the long-running experiments on the famous African Greys parrots Alex, Griffin, and Arthur,[4] can be criticised for taking animals from the wild or from zoo shops to live their lives under laboratory conditions, far from the lives they should have had. Alex was bought in a zoo shop, and the anoles were taken from the wild in Puerto Rico (there was no ethical discussion in the paper about their fate) (Leal and Powell, 2011).[5] Although the experiments and training of Alex, Griffin, and Arthur may generate knowledge that eventually may be an advantage to other African Greys, as it may result in an elevated consciousness about the cognitive skills of these and other parrots, from these specific individuals' perspective it is reasonable to claim that they are victims of the abuse of power of scientists, although the parrots also may benefit from mutual relationships with the scientists. This issue is relevant for the discussion of individual versus species justice, to which I return towards the end of the chapter. A crueller example of how even the science industry is part of the wildlife trade is provided by the fact that 10 per cent of the 10,000 primates used in test research in European Union (EU) laboratories per year are wild-caught (ADI, 2009; see also Zimmerman, 2003), thus providing another example of exploitation, which of course extends to all animals suffering in vivisection, whether wild-caught or not.

Wildlife trafficking in Norway: a case study

In addition to being protected under the CITES Convention, Norwegian fauna and flora are also protected under the Law[6] of Biological Diversity and the Bern Convention.[7] My data reveal that animals are smuggled into Norway, and they link Norway to the global wildlife market.[8]

So far I have interviewed four police officers and police lawyers in the Eco-crime (hereafter Eco-crime) section of the police department, which is the section in charge of investigating crimes against the environment in Norway. I have interviewed one veterinarian at Oslo Airport, who is responsible for animal welfare at the airport, which includes those animals who are transported in and out of the country. Three persons have been interviewed in the Customs Directorate, in addition to an operative customs officer.

On the offender side, I have interviewed five persons who keep and have smuggled illegal reptiles into Norway. I have also interviewed a person at the Directorate for Nature Management (DN), which is the institution in charge of CITES in Norway. He also happens to be central in the CITES organisation. I have also had several telephone interviews with environmental crime coordinators in the police department in different border districts.

I have also analysed verdicts for breaches of the CITES regulations[9] and Customs laws, which regulate what can be taken in and out of the country. I have been given access to 43 confiscation reports from Customs from CITES cases, which describe under what circumstances persons are apprehended for smuggling animals and animal parts into Norway. Finally, I have studied several reptile and parrot websites. The interviews lasted approximately two hours, were recorded, and have been transcribed.[10]

Findings: the tip of the iceberg?

Interviews with the police, the border veterinarian, and Customs indicate that the control of animals and animal derivates coming into the country is too weak. Norway is located at the very top of Europe; it has a very long coastline, a long border with Sweden, and also a border with Russia in the north, as well as being a Schengen Agreement and European Union (EU)[11] frontier state. To stop the smuggling of animals into Norway, everything and everyone coming into the country would have to be examined. As Alacs and Georges (2007) and Yi-Ming et al. (2000) state in the cases of Australia and China, respectively (see also Wellsmith, 2010: 135–136), this would be an impossible task, as it would include control of container shipments, post, and keeping permanent vigilance over all the country's Custom posts. People, cars, containers, etc, are rather randomly controlled, and are controlled on suspicion.

According to the DN, the easiest way to smuggle endangered animals and products from endangered species into Norway is through the forwarding companies that transport merchandise internationally, and Customs and Eco-crime are therefore requested by DN to focus on these. For example, containers arrive in Oslo Harbour with stuffed big cats and crocodiles. Whether these animals are legally exported and imported can be hard to establish, and containers will seldom be opened. The Customs Directorate has had coordinated operations with the Norwegian Food Safety Authority [Mattilsynet] (ironically, in charge of animal welfare) and Eco-crime. For example, one operation concentrated on meat coming in from outside the EU. During this operation, meat from Central Africa was carried in by a man into Oslo Airport, where it was confiscated. The meat may well have come from illegal bush-meat slaughter (Boekhout van Solinge, 2008a; Broad, Mulline and Dileys, 2005). However, neither the Customs Directorate nor the Mattilsynet (according to my interviewees) possesses the necessary resources to do a DNA analysis of the confiscated meat and thereby determine what kind of animal the meat is from. The meat was confiscated for food safety reasons, not because it could have come from CITES-listed animals. Confiscated animal parts are *not* routinely sent to analysis.

In 2011, according to Customs confiscation statistics, on 20 occasions, a total of 331 live animals were confiscated, of which 2 were birds, 74 were eggs, 240 were reptiles, and 15 were mammals. In 2010, a total of 66 animals were confiscated on 27 occasions, of which 14 were birds, 14 were mammals, and 38 were reptiles. In 2009, 541 live animals were confiscated on 40 occasions, of which 397 were reptiles, 24 were birds, 22 were eggs, and 88 were mammals. In 2008, 112 animals were confiscated on 38 occasions, of which 23 were reptiles, 51 were birds, and 27 live mammals. Though many of these confiscated animals were reptiles, fewer animals that are CITES-listed were confiscated. In

2011, there were in total 131 CITES confiscations, in which 7 reptiles, 40 birds, and 1 mammal were confiscated. In 2010, there were 149 CITES confiscations, in which 5 reptiles were confiscated (as part of total 15031 CITES confiscations). In 2009, 7 reptiles were confiscated, while in 2008, 6 CITES-listed reptiles and 51 CITES-listed birds were confiscated. According to the veterinarian at Oslo Airport, however, they confiscate turtles weekly, which may indicate that even the statistics from Customs show only part of the total number of confiscations. This impression was confirmed by my interviewee in CITES and DN. Many people carry a turtle in their pocket. At one occasion, somebody evidently got cold feet at the airport and the little turtle was left to walk in through the green zone all by herself.

In the Eco-crime police unit, they admit that they do not have an overview of the extent to which CITES-listed animals are smuggled into Norway – for example, because not all confiscations will result in police reports. The Mattilsynet may not report it to the police, because they consider this the responsibility of Customs, while the environmental crime coordinators of the police may not prioritise the incident. In addition, the CITES cases are not consistently coded, so they may be hard to detect in the system, and this also applies for CITES verdicts, as they are coded under different laws. For example, when Germans carry brooding machines into Norway in order to carry eagle and falcon eggs out, these cases can be coded under the Bern Convention regulations. These smugglers are often part of rings of collectors, as when a man in Finland recently was arrested who had more than 10,000 eggs and hundreds of stuffed birds from endangered species in his possession, many of which, according to the Finnish police, were abducted from Norway from protected areas and national nature reservations (Rapp, 2012).

The problem of insufficient skill and coordination, and scarce resources

Findings in the international literature imply that there is a problem connected to CITES, that a goal of the Convention is to regulate trade in threatened species, which implies that those who shall control the trade, not the least police and Customs, must possess certain qualifications if control is to be successful (Warchol et al., 2003; Zimmerman, 2003). A Customs officer must be able to distinguish an individual animal from CITES-listed species from one that is not listed. The literature also sets demands regarding the knowledge of what kind of instrument CITES is. Forgery of documents and corruption are other problems (Warchol et al., 2003).

Frontline law enforcement agencies in Norway face similar problems in identifying protected species, but in Norway, Customs have support in the DN, which is the institution that, with the support of the Norwegian Veterinary Institute, identifies the species to which an individual animal belongs. An encounter with the Mattilsynet is the instance that, through their border veterinarian, someone first meets the animals that are smuggled into Norway, once Customs uncovers them. One of my interviewees thus told me that situations could occur in which she would be called because the Customs officer feared for animals' lives. Once she was called because a man from Thailand was attempting to smuggle birds into the country. They were stuffed into a bottle, in a row, about every second bird was dead, which can serve as an example of the fact that up to 75 to 90 per cent of birds that are trafficked die during transport (Guzmán et al., 2007). She also said that she would agree that the situation was critical; paperwork, that is the verification of export and import permits, was done so fast that they were not adequately checked, meaning in reality that she, as a veterinarian, rather than Customs, would verify the documents and thus accept the import of the animal(s). For example, she claimed, very often fish are taken hastily through Customs, because many are already dead and dying. And, as she said, she could not distinguish a forged export permit from a genuine one. This would imply that the veterinarian, rather than Customs, is in fact giving the 'go ahead' for illegal import of CITES animals. This, however, was disputed by Customs at the airport, but they nonetheless admitted that verification of the documents represented a challenge they are not equipped to handle. The Customs officer also complained of the lack of cooperation between these control agencies.

The interviews with the control agencies indicate that Norway's control and law enforcement in this field are still relatively underdeveloped. It seems like in each organisation there are a few very dedicated persons who are left with the responsibility for the enforcement in their sections, thus leaving the total control and enforcement vulnerable. Lack of dedication from persons who have the responsibility in the border districts, and scarce resources and low priorities in the investigation, may also result in underenforcement, and as a result, cases fail to be reported to the Eco-crime police section and no one is prosecuted. Cooperation among the different control groups is not sufficiently developed and prioritised.

It can also create frustration if the police abstain from prosecution in cases in which Customs have confiscated animals and have reported animal trafficking, even though they are 'served the case on a tray', as one interviewee put it. Customs efforts are also guided by cooperation with the International Police and Customs. For example, Norway took part in Operation TRAM in February 2010, a joint operation involving police, Customs, and other specialised units in a number of countries that was directed at CITES-protected species used in Asian medicine. However, participation in Operation RAMP, directed at the illegal trade and trafficking in reptiles and amphibians, which took place in September and October 2010, in which 51 countries from five continents took part, was not prioritised in Norway (Brørby, 2010). This may seem like the wrong decision, given the extent of the reptile trade worldwide (Herbig, 2009; Alacs and Georges, 2007); estimates indicating that there are at least 100,000 illegal reptiles in Norway (Sunde, 2010) imply that Norway is part of the trade. In the next section, I will show how some of those who have been caught smuggling animals and animal parts are punished in Norway.

Verdicts

I have gone through seven verdicts for CITES cases. These seven are all that Eco-crime was able to detect at a given moment (February 2011), but it is unlikely that these are the total number of CITES verdicts (and fines). The fact that the cases are coded under different laws and regulations makes it difficult to trace them in the statistics. Compared to the number of confiscation reports appearing in Customs statistics, these verdicts are very few. This can, however, also indicate that many cases are never reported to the police, or that no one is convicted or fined.

Three of the verdicts concerned the same case, from the First Instance Court, Forhørsretten [Magistrates Court] to Courts of Appeal, Lagmansretten and Høyesterett [The Supreme Court]. Two other cases ended with a fine only, and therefore provide little information about the cases, but one of them concerns the smuggling of a large number of Diamond sturgeon, Sterlet sturgeon, and Siberian sturgeon, which are CITES-listed species whose individuals are killed for the production of caviar. The market for caviar is huge, and it is endangering the sturgeon. For example, according to Cowdry (in Schneider, 2008: 281), US $25 million worth of caviar left the United Arab Emirates illegally for destinations in the United Kingdom and the United States. This illegal act thus connects Norway to a harmful global market (see also Dickson,

2005). Still, the offender got only a fine of 10,000 NOK (Norwegian krone)[12] and 12,000 NOK confiscations of the economic gains of the crime in favour of the state.

Two other cases are particularly interesting because of the motives and the great numbers of animals who were harmed and killed. In the Supreme Court case, the convicted person had sent a box by mail containing 271 objects with feathers and animal parts from two margays (a spotted cat native to Central and South America), between 10 and 20 parrots, and two kvelerslange/Python, all CITES-listed. The final Supreme Court sentence was 45 days of suspended prison. This can be regarded as very lenient punishment, and it corresponds to the situation in other countries (Lowther, Cook and Roberts, 2002). Yet the judge argued that it was important to punish such crimes *severely* to deter others. The judge further argued that the convicted should not be burdened by the delay in the court system; it took three years from the time that the crime took place to the final verdict.

The other case of particular interest was against a man who was sentenced to prison for 120 days, in addition to confiscation of economic gains of the crime of 34,000 NOK, and the prohibition from being in possession of non-Norwegian species for five years. The offender was convicted in 1998 for 11 occasions of being guilty of smuggling 31 birds listed in CITES 2, predominantly parrots, eight macaws from the CITES 1 list, 2 boas (CITES 2 listed), and 50 turtles, in addition to a number of other animals and birds, in order to sell them in Norway. The accused said in court that he had made such trips more than 20 times to buy birds in Sweden and the Netherlands. My data indicate that this is likely the same man who was stopped on the Norwegian-Swedish border *Svinesund* with 8 African Grey parrots (CITES Appendix 2 list) and a lot of spirits, wine, and beer in November 2011. This man had been in contact with DN, and during this conversation he admitted to have no plans to stop smuggling, because it is so lucrative. My interviewee in DN/CITES said there was a high probability that these birds were wild-caught, as wild-caught parrots in the Netherlands are far cheaper than those that are locally bred. Both these cases show that economic gains by the sale of CITES-listed animals is the motive behind the crimes, and that Norway, although being located on the periphery of Europe, is part of the international wildlife trade and markets. Given the fatality percentages for wild-caught parrots, it is possible to imagine the total costs in terms of bird lives for just one of the mentioned acts of illegal trafficking into Norway. If the total of 31 birds were smuggled in, these may be the surviving 10 per cent of 310 birds, which may have been taken from their habitat by pure force.

As mentioned, there are a number of laws and regulations that apply when somebody traffics exotic species into Norway without the required CITES permits. These import permits are authorised in Norway by DN on the basis of presented export permits, according to CITES regulations. The verdicts I have analysed show that people have been convicted under (among others) Customs law, regulations about prohibition of introduction of animals and contagious objects, regulations about the trade in endangered species (CITES), the Animal Welfare Act, and a regulation about animal welfare in trade.[13]

Based on the few verdicts I have analysed, it is hard to draw general conclusions, yet I find it unlikely that the verdicts will work to deter anybody from committing such crimes (see also Alacs and Georges, 2007: 155; Yi-Ming et al., 2000; Lowther et al., 2002). My data further suggest that smuggling exotic animals for sale on the pet market for some is combined with other smuggling of liquor and cigarettes, which is also lucrative in Norway due to the high price of these goods.

Other typical confiscation reports reveal that people smuggle reptiles for their own use, as pets, or buy a turtle in a market in Turkey and put it in their luggage, or tourists carry parts of dead animals, either transformed into belts and souvenirs, or simply as heads and pelts (for example, crocodile heads, bear and wolf pelts). In examining post packages, Customs has found products in which animals are ingredients, such as patches containing leopard and tiger bone and musk. In a random control, Customs revealed a primate head and the suspect woman said this was a test, she wanted to see what she could get for it, and then establish import on a larger scale, this despite the fact that most primates are CITES 1 listed. Those who traffic animals as part of business seem calculate that sometimes they will get caught, but that the gains outweigh the costs, in terms of a fine.

The confiscation reports further indicate that those who are caught in the act of smuggling an animal or part of one usually are aware that they are committing a crime. In the case of reptile smuggling, the practice of using two cars when smuggling animals in from Sweden is frequent; the second car, with the animal(s), will follow if the first one crosses the border unhindered by Customs.

The reptile keepers and smugglers

My interviewees who keep reptiles have often smuggled them to Norway themselves. The reptiles are bought in reptile fair markets – the Terraristika Fair in Hamm, Germany,[14] which is notorious for its exposure of wild-caught animals, is often mentioned – or they buy them in zoo shops in Denmark and Sweden. One should assume that animals that are for sale in pet shops would be locally bred, yet my interviewees are convinced that many of these reptiles are wild-caught. They establish this by assessing the general health of the animals that are exposed and the parasites they seemingly have. Wild-caught animals often have parasites. Zoo chains' involvement in the illegal trafficking in animals has been previously documented in the United States, where the Pet shop chains Petco, Petsmart, and Petland bought, trafficked, and sold animals. Hundreds of thousands of animals died before they reached their buyers, as they were trafficked and kept under horrible conditions.[15]

My interviewees who keep reptiles are aware that this is illegal. These interviews touch traditional, criminological topics related to stigmatisation, social exclusion, and criminalisation. The reptile keepers experience a situation in which they, as a consequence of hiding their illegal animals, often isolate themselves socially from people who do not share their interest and situation, and as a result also socialise all the more with other reptile keepers, partly forming a subculture. They will not open the door if the door bell rings and live with the constant fear of being informed on or caught and having their animals taken away from them. As a result of keeping reptiles, their social network is reduced. Those who have been caught are afraid that this may ruin their future careers.

The Norwegian reptile interest group thus argues for legalising reptiles in Norway and has developed a 'positive list' of 31 reptile species that can be physically handled. Mattilsynet is now considering this list and has ordered a risk analysis from the Scientific Committee for Food Safety to assess the risk of bad animal welfare and infections. As many of the animals on the list are CITES-listed, the committee suggests that the animals should be bred in captivity.

Criminalisation versus legalisation and regulation

Norway and Iceland are the only countries in Europe that prohibit the sale and keeping of reptiles. Exceptions can be made in Norway for persons who are allergic to other animals, who may apply to the Norwegian Food Safety Authority for permission to keep a tortoise. This unique situation also makes Norway an interesting case for the discussion of implications regarding parallel markets and regulation of trade, rather than criminalisation, as these present specific challenges in terms of control and law enforcement. Traffickers in endangered species can declare their products to be from different species, especially where there are a number of species of similar appearance (Hayman and Brack, 2002: 20). They can also claim that illegally caught animals have been bred in captivity (Wright, 2011: 337). Though CITES may have contributed to reduced deaths and species extinction, it is also documented that CITES regulation entails significant problems in terms of verification of import and export certificates, and in the skills needed for separating genuine from forged certificates (Warchol et al., 2003; Rosen and Smith, 2010). Consequently, with the existence of parallel licit and illicit markets, animals that should be protected under CITES may be 'laundered', and neither buyer nor control agencies may be aware that a wild-caught, illegal animal is the victim of abduction, trafficking, and trade (Lowther et al., 2002). Corruption is another problem (Schmidt, 2004), which also points to the great difficulties in many states' will and capacity to effectively enforce CITES, according to my DN/CITES interviewee. He estimated that at least half of the 175 parties to CITES do not efficiently enforce the regulation and prosecute offenders. Political instability and warlike situations, for example in Central Africa, is one reason.

Legal trade may encourage illegal trade: 'Illegal trade is rightly regarded as both difficult to control and more likely to lead to unsustainable harvesting than legal harvesting, and there are plenty of examples where illegal trade has flourished under the cover of legal trade' (Hutton and Webb, 2005: 109). Still, in taking the offender's perspective in the reptile case of Norway, one may question whether reptile keepers should be criminalised any more than parrot keepers, and why they should be criminalised any more than reptile keepers in Sweden and Denmark. Given the costs of criminalisation of people who qualitatively do nothing else than keep other 'pets', their practices could also be legalised. Such logic may also be one reason why keeping reptiles is legal in most countries. When it is illegal to keep reptiles, it may also be difficult to provide them with the necessary veterinary assistance.

However, one evil should not be added to another because of the presumed legitimacy of the first, or for other reasons. From the animals' perspective, other animals' suffering does not ameliorate their own suffering. Legalising the reptile market in Norway would likely produce an increase in the practice of keeping reptiles, and consequently an increase in the trade. It would challenge Customs officers in separating the legal from the endangered CITES-listed ones, a problem that was highlighted in interviews. Although it is possible that locally bred individuals from Norway and Denmark may cover much of the demand, as wild-caught animals generally are cheaper than hand-bred ones, a legalisation of the Norwegian market could also entail an increase in the trafficking and trade of protected species (for example, see Herbig, 2010). Undoubtedly it would entail an increase in abuse, suffering, and death for many individual animals. On the other hand, a continued prohibition can also reinforce the value of some species for collectors, as it has been documented that the rarer the remaining individuals of a species, the higher the price (Kendall and Reeve, 2002; Zimmerman, 2003; Sollund, 2011).

Species justice, ecological justice, and individual rights

Rob White (2007, 2008, 2011 and Chapter 1 of this volume) highlights the concepts *ecological justice, species justice, and environmental justice*. I find these useful in discussing the harmful effects of the wildlife trade. *Ecological justice* implies that humans are but one part of complex ecosystems, which should be preserved for their own sake. A concern is for planetary well-being and the rights of other species to live free from abuse and torture (White, 2007: 38 and 2011: 23). *Environmental justice* is used as a prolongation of human rights or social rights to enhance the quality of human life, now and for the future (White, 2007: 38–39; 2008: 18–21 and 2011: 23). *Species justice* implies that harm is seen in relation to the place that nonhuman species have in their environments and their intrinsic right not to suffer from abuse, whether one to one or institutionalised, or as consequence of human action that harms habitats, the climate, and the environment on a global scale. As I read White, this would imply taking an ecocentric perspective (Halsey and White, 1998).

A question that I think fails to be answered is how, precisely, species justice relates to individual rights and individual justice. Can or should species justice – in terms of species survival – be prioritised over the individual rights of those belonging to the species? Or, on the other hand, can species justice only be accomplished if the individual members of a species are not mistreated, exploited, and killed?

Another question is why, as I read White, environmental rights are an extension of *human* rights (only). Why shouldn't environmental rights – understood as rights of the environment – be justly, equally distributed to human *and* nonhuman species? From a biocentric perspective, one could claim that nonhuman species are more important in preserving ecosystems than humans, as humans only live on and by other ecosystems, not as an integrated part of them. Can species justice be accomplished at all, unless all individuals of all species have equal rights?

In his elaboration of *ecological citizenship* (White, 2011: 23, 147–148), White states that the concept incorporates 'the key concerns of environmental justice, ecological justice and species justice'. I read him as implying that ecological citizenship also foremost applies to *humans*, as humans are perceived as having the responsibility for other species, and are, and should be, positioned *above* them, as some sort of gardener, implying an ecocentric perspective (Halsey and White, 1998). White (2007: 42) is quite specific about *not* granting other species equal moral rights, when he in his argument against biocentrism compares a mosquito to a human, thus concluding that biocentrism can entail misanthrocentrism, and perceptions that are 'morally repugnant and politically suspect'. It is a question however why, instead of a mosquito, White does not compare a human to a nonhuman animal such as a mammal, in discussing why humans should have rights that nonhuman animals are not entitled to (see Regan, 1983 for this discussion). As Nussbaum states: 'The touchstone should be respectful consideration of the species norm of flourishing and a respectful attention to the capacities of the individual' (Nussbaum, 2006: 378). Nonhuman animals' important roles in ecosystems enhance why they, particularly, should be secured ecological citizenship and rights to their habitats. A consequence of ecological citizenship and ecological justice would be that individuals of all species should have environmental rights, and consequently, should have both individual rights and species-related rights.

This brings me to a discussion of the rationale behind CITES, in light of ecological justice and species justice. The abduction, trafficking of nonhuman animals, and theriocide (the killing of nonhuman animals by humans; Beirne, 2009a, b) is legitimated through CITES as nonhumans are consistently regarded as 'natural resources' which can be 'harvested' for human benefit (Sollund, 2011). To 'over'-exploit them, though, is unacceptable, principally because this will eventually harm humans, and can even, when CITES is incorporated in national legislation, be against the law, or at least against regulations, as in the Norwegian case. Whether illegal or not, the abduction and theriocide of nonhumans that CITES indirectly encourages has tremendous consequences in terms of individual (and species) suffering. When species are driven to extinction, this not only harms the individuals who form the species, but also the ecosystems to which the individuals belong.

When the ecosystem to which the parrot belongs is drained, this is a breach of ecological justice, and when the species suffer as a consequence of annual 'harvesting' by humans, as occurs in 'La loreada', this is also a breach of species justice, as it is when species go extinct due to loss of habitat because of humans' logging activities (Halsey and White, 1998). The abduction and/or killing of parrots is at least a breach of individual rights of parrots to live unharmed by humans.

We can thus distinguish between *individual rights,* which should be but are not necessarily connected to *species-specific rights,* meaning those rights pertaining to the specific needs of one particular species, for example to dig, crawl, fly, migrate, run, or search for food. It is a breach of both species-specific rights and individual rights, for example, when the last remaining individuals are abducted from their habitats to be bred in captivity, deprived of their freedom and habitat. A species can survive in a zoo, but this survival can be in contrast to individual and species-specific rights. From this we can deduce that even if the species survives, and the species at least from one dimension is given justice, this may be inconsistent with ecological justice. Species survival can be but is not necessarily connected to the right of the species to continue its existence in its natural environment and consequently, to species justice. It is also a question whether species justice can at all be fulfilled if this is only accomplished at the cost of the rights of individuals of that species, for can a species be *more* than the individuals who form it? A species without individuals is merely an analytical category (Svärd, 2008).

Conclusion

Illegal animal trafficking and trade is tremendously harmful and is expanding. According to my data, Norway is part of the trade and can serve as one example of the multitude of variations and motives that exist for partaking in the trade – for example, the pet trade; trophy hunting and trophy trade; collecting; and animal parts used as adornments and souvenirs.

Based on my data, it seems like Norwegian control agencies do not have an overview of the illegal animal trade in Norway. Nor do they have sufficient knowledge, and so far control and enforcement seems poorly prioritised. The control is vulnerable because it depends on the efforts of a few dedicated persons who possess much of the knowledge on the field. The level of punishment can hardly deter anybody, as the gains of the trade are significant.

Both the legal and the illegal trade must be seen from the perspective of harm and (in)justice, focusing on the real victims of the trade, the animals themselves. Animals, whether belonging to endangered species or not, must be recognised as victims when humans abuse their power and violently abduct them from their natural habitats, after which time these nonhuman animals, if they survive, will live the rest of their lives as eternal captives.

Notes

1. The concept 'wildlife' may be regarded as alienating and anthropocentric (White, 2001; Sollund, 2011), in seeing animals as a mass and by othering them by regarding them as 'wild', and also by implying that they are resources for human use. For simplicity I will still use the word in the chapter, though with a meaning that does not include 'domesticated nonhuman animals'.
2. See the CITES webpage: http://www.cites.org/
3. The total number of private birds, (not 'production birds', such as hens and geese) is estimated to be 414,000 (Mejdell, 2004).
4. See the Alex foundation website: http://www.alexfoundation.org/index2.htm
5. The readers are not told whether the anoles were returned to their habitats after the experiments.
6. Lov om forvaltning av naturens mangfold (naturmangfoldloven) [Law about the management of biological diversity], LOV-2009-06-19-100. http://www.lovdata.no/all/nl-20090619-100.html
7. See the Convention on the Conservation of European Wildlife and Natural Habitats, CETS No.: 104 at: http://conventions.coe.int/Treaty/Commun/QueVoulezVous.asp?NT=104&CM=8&DF=09percent2F10percent2F01&CL=ENG
8. Many animals, especially dogs who are not CITES-listed, are smuggled into Norway. The smuggling of these animals, despite the harms the dogs suffer because of the breeding and trade, is beyond the scope of this chapter, as I here focus on those animals which are so-called 'wild', not those domesticated by humans.
9. FOR 2002-11-15 nr 1276: Forskrift til gjennomføring av konvensjon 3. mars 1973 om internasjonal handel med truede arter av vill flora og fauna (CITES). This regulation is currently under change and a proposition for a new CITES regulation has been sent from the Directorate for Nature Management (DN) to the Ministry of Environmental Affairs, and will likely be accepted by the National Assembly within 2013. This, according to the DN/CITES interviewee.
10. The project has been reported to and is accepted by Norwegian Social Science Data Services and follows ethical research guidelines.
11. Norway is not part of the European Union, but takes part in most EU regulations and agreements through the European Economic Area (EEA).
12. At the time of writing about 6 NOK is equivalent to 1 USD.
13. LOV-2007-12-21-119, Tolloven, FOR-1991-07-02-507, Forskrifter om forbud mot innførsel av dyr og smittebærende gjenstander, FOR 2002-11-15 nr 127,Forskrift til gjennomføring av konvensjon 3. mars 1973 om internasjonal

handel med truede arter av vill flora og fauna (CITES) 6, LOV-1997-06-06-32, Lov om innførsel og utførselregulering, LOV-2009-06-19-97 Lov om dyrevelferd, and FOR 1985-01-10 nr 17: Forskrift om velferd for dyr ved ervervsmessig omsetning.

14. See: http://www.youtube.com/watch?v=tLAJR_dlL28
15. See the A PETA Undercover Investigation at http://features.peta.org/pettrade/default.asp

References

Animal Defenders International (AND) (2009) *Primate Testing in Europe. A Report on the Use of Primates in a Typical European Commercial Testing Laboratory.* Available at http://www.savetheprimates.org/files/stpreporten.pdf. Accessed 30 June 2012.

Agnew, R. (1998) 'The causes of animal abuse: a social-psychological analysis', *Theoretical Criminology*, 2(2): 177–209.

Alacs, E. and Georges, A. (2008) 'Wildlife across our borders: a review of the illegal trade in Australia', *Australian Journal of Forensic Sciences* 40(2): 147–160.

Beirne, P. (1999) 'For a nonspeciesist criminology. Animal abuse as an object of study', *Criminology*, 37(1): 117–149.

Beirne, P. (2009a) 'Animal rights. Animal abuse and green criminology', in P. Beirne and N. South (eds) *Issues in Green Criminology.* Devon: Willan, pp. 55–87.

Beirne, P. (2009b) *Confronting Animal Abuse.* Lanham, MD: Rowman & Littlefield Publishers.

Boekhout van Solinge, T. (2008a) 'Crime conflicts and ecology in Africa', in R. Sollund (ed.) *Global Harms Ecological Crime and Speciesism.* New York: Nova Science Publishers.

Boekhout van Solinge, T. (2008b) The land of the orang-utan and the birds of paradise under threat', in R. Sollund (ed.) *Global Harms Ecological Crime and Speciesism.* New York: Nova Science Publishers.

Broad, S., Mulline T. and Dileys, R. (2005) 'The nature and extent of legal and illegal trade in wildlife', in S. Oldfield (ed.) *The Trade in Wildlife. Regulation for Conservation.* London: Earthscan.

Brørby, T. (2010) '"Operasjon RAMP". Internasjonal aksjon mot illegal handel av reptiler og amfibier', *Miljøkrim*, 1.

CITES Convention. Available at http://www.cites.org/eng/disc/text.shtml. Accessed 14 June 2011.

Dickson, B. (2005) 'What is the goal of regulating wildlife trade? Is regulation a good way to achieve the goal?' In S. Oldfield (ed.) *The Trade in Wildlife. Regulation for Conservation.* London: Earthscan, pp. 23–33.

Flynn, C. (2011) 'Examining the links between animal abuse and human violence', *Crime, Law and Social Change*, 55(5): 453–468.

Gonzales, J. A. (2003) 'Harvesting, local trade, and conservation of parrots in the Northeastern Peruvian Amazon', *Biological Conservation*, 114: 437–446.

Guzmán, J. C. C., Saldaña, M. E. S., Teyeliz, A. C., Grosselet, M. and Gamez, J. S. (2007) *The Illegal Parrot Trade in Mexico. Defenders of Wildlife.* Available at www.defenders.org/mexicanparrot. Accessed 25 June 2012.

Halsey, M. and White, R. (1998) 'Crime, ecophilosophy and environmental harm', *Theoretical criminology*, 2(3): 345–371.

Herbig, J. (2010) 'The illegal reptile trade as a form of conservation crime: a South African criminological investigation', in R. White (ed.), *Global Environmental Harm: Criminological Perspectives*. Devon: Willan Publishing.

Herrera, M. and Hennessey, B. (2007) 'Quantifying the illegal trade in Santa Cruz de la Sierra, Bolivia, with emphasis on threatened species', *Bird Conservation International*, 17: 295–300.

Hutton, J. and Dickson, B. (2000) (eds) *Endangered Species Threatened Convention*. London: Earthscan Publications.

Hutton, J. and Webb, G. (2005) 'Crocodiles: legal trade snaps back', in S. Oldfield (ed.) *The Trade in Wildlife. Regulation for Conservation*. London: Earthscan, pp. 11–120.

IFAW (International Fund for Animal Welfare) (2008) *Killing with keystrokes. An investigation of the illegal wildlife trade on the World Wide Web*. Available at http://www.ifaw.org/Publications/Program_Publications/Wildlife_Trade/Campaign_Scientific_Publications/asset_upload_file64_12456.pdf. Accessed 14 June 2011.

Leal, M. and Powell, B. J. (2011) 'Behavioural flexibility and problem-solving in a tropical lizard', *Biology Letters*, (8): 28–30.

Lemieux, A. M. and Clarke, R. (2009) 'The international ban on ivory sales and its effects on elephant poaching in Africa', *British Journal of Criminology*, 49(4): 451–471.

Lowther, J., Cook, D. and Roberts, M. (2002) *Crime and Punishment in the Wildlife Trade* WWF/Traffic. Regional research institute. West Midlands, UK: University of Wolverhampton.

Maher, J. and Pierpoint, H. (2011) 'Friends, status symbols and weapons: the use of dogs by youth groups and youth gangs', *Crime, Law and Social Change*, 55(5): 405–420.

Mejdell, C. (2004) 'Katt eller kanarifugl?– 37 percent av norske hjem holder dyr' [Cat or Canary? 37 percent of Norwegian homes keep animals] In *Norsk veterinærtidsskrift* 7(116): 536–537.

Michaels, A. (2002) 'Parrot smuggling. Still a global problem', *Animal Welfare Institute. Quarterly*. Fall 2002, 51(4). Available at http://labanimals.awionline.org/pubs/Quarterly/fall02/parrot.htm. Accessed 24 August 2012.

Nibert, D. (2002) *Animal Rights Human Rights*. Oxford: Rowman & Littlefield publishers.

Nussbaum, M. C. (2006) *Frontiers of Justice. Disability, Nationality, Species Membership*. Harvard: The Belknap Press of Harvard University Press.

Pepperberg, I. (1999) *The Alex Studies. Cognitive and Communicative Abilities of Grey Parrots*. Cambridge, MA.: Harvard University Press.

Pires, S. and Clarke, R. V. (2011a) 'Sequential foresting, itinerant fences and parrot poaching in Bolivia', *British Journal of Criminology*, 51(2): 314–335.

Pires, S. and Clarke, R. V. (2011b) 'Are parrots CRAVED? An analysis of parrot poaching in Mexico', *Journal of Research in Crime and Delinquency*, 49(1): 122–146. doi:10.1177/0022427810397950.

Pires, S. and Moreto, W. B. (2011) 'Preventing wildlife crimes: solutions that can overcome the "tragedy of the commons"', *European Journal of Crime Policy*, 17(2): 101–123.

Rapp, O. M. (2012) Ulovlige egg er big business [Illegal eggs are big business], *Aftenposten*, 10 June, p. 8.

Regan, T. (1983) *The Case for Animal Rights*. Berkeley: Berkeley University Press.

Reeve, R. (2002) *Policing International Trade in Endangered Species. The CITES Treaty and Compliance*. London: Earthscan.

Rivalan, P., Delmas, V., Angulo, E., Bull, L. S., Hall, R. J., Courchamp, F., Rosser, A. M. and Leader-Williams, N. (2007) 'Can bans stimulate wildlife trade'? *Nature*, 447: 529–530.

Rosen, G. E. and Smith, K. (2010) 'Summarizing the evidence on the international trade in illegal wildlife', *Ecohealth*, 7(1): 24–32.

Schmidt, C. (2004) 'Environmental crimes: profiting at the Earth's expense', *Environmental Health Perspectives*, 112(2): 96–103.

Schneider, J. L. (2008) 'Reducing the illicit trade in endangered wildlife: the market reduction approach', *Journal of Contemporary Criminal Justice*, 24(3): 274–295.

Sollund, R. (2008) 'Causes for speciesism: difference, distance and denial', in R. Sollund (ed.) *Global Harms Ecological Crime and Speciesism*. New York: Nova science publishers, pp. 109–131.

Sollund, R. (2011) 'Expressions of speciesism: the effects of keeping companion animals on animal abuse, animal trafficking and species decline', *Crime, Law and Social Change*, 55(5): 437–451.

South, N. and Wyatt, T. (2011) 'Comparing illicit trades in wildlife and drugs: an exploratory study', *Deviant Behavior*, 32(6): 538–561.

Stoett, P. (2002) 'The international regulation of trade in wildlife: Institutional and normative considerations' International Environmental agreements. *Politics, Law and Economics*, 2(2): 195–210.

Sunde, S. (2010) *Ett av tusen ulovlige reptiler i Norge blir beslaglagt?* [One of a thousand reptiles confiscated?] Miljøkrim. Tidsskriftet for miljøkriminalitet. Available at http://www.okokrim.no/miljokrim/nor/tidligere_utgaver/3_desember_2010/artikler/ett-av-tusen-ulovlige-reptiler-i-norge-blir-beslaglagt. Accessed 30 June 2012.

Svärd, P. A. (2008) 'Protecting the animals? An abolitionist critique of animal welfarism and green speciesism', in R. Sollund (ed.) *Global Harms: Ecological Crime and Speciesism*. New York: Nova science publishers.

Traffic (2008) *What's Driving the Wildlife Trade? A Review of Expert Opinion on Economic and Social Drivers of the Wildlife Trade and Trade Control Efforts in Cambodia, Indonesia, Lao PDR and Vietnam*. East Asia and Pacific Region Sustainable Development Discussion Papers. East Asia and Pacific Region Sustainable Development Department, World Bank, Washington, DC. Available at http://siteresources.worldbank.org/INTEASTASIAPACIFIC/Resources/226262-1223319129600/wildlife_foreword_terms_acknowledgement.pdf. Accessed 4 May 2011.

Warchol, G. L., Zupan, L. and Clarke, W. (2003) 'Transnational criminality: an analysis of the illegal wildlife market in Southern Africa', *International Criminal Justice Review*, 13(1): 1–26.

Wasser, S. K., Clark, B. and Laurie, C. (2009) 'The ivory trail', *Scientific American*, 301: 68–76.

Wellsmith, M. (2010) 'The applicability of crime prevention to problems of environmental harm: a consideration of illicit trade in endangered species', in

R. White (ed.) *Global Environmental Harm. Criminological Perspectives*. Devon: Willan, pp. 132–150.

Wellsmith, M. (2011) 'Wildlife crime: the problems of enforcement', *European Journal of Crime Policy*, 17(2): 125–148.

Weston, M. K. and Memon, M. A (2009) 'The illegal parrot trade in Latin America and its consequences to parrot nutrition, health and conservation', *Bird Populations*, 9: 76–83.

White, R. (2007) 'Green criminology and the pursuit of social and ecological justice', in P. Beirne and N. South (eds) *Issues in Green Criminology*. Devon: Willan Publishing.

White, R. (2008) *Crimes against Nature. Environmental Criminology and Ecological Justice*. Devon: Willan Publishing.

White, R. (2011) *Transnational, Environmental Crime. Toward an Eco-global Criminology*. London: Routledge.

Wilson-Wilde, L. (2010) 'Wildlife crime: a global problem. Commentary', *Forensic Sci Med Pathol*, 6(201): 221–222.

Wright, G. (2011) 'Conceptualising and combating transnational organised crime', *Trends in Organized Crime*, 14(4): 332–346.

Wright, T. et al. (24 co authors) (2001) 'Nest poaching in neotropical parrots,' *Conservation Biology*, 15(3): 710–720.

Wyatt, T. (2009) 'Exploring the organization in Russia far east's illegal wildlife trade: two case studies of the illegal fur and illegal falcon trades', *Global Crime*, 10(1–2): 144–154.

Wyatt, T. (2011) 'Illegal raptor trade in the Russian Federation', *Contemporary Justice Review*, 14(2): 103–123.

Yi-Ming, L., Xenxiang, G., Xinhai, L., Sung, W. and Niemelä, J. (2000) 'Illegal wildlife trade in the Himalayan region of China', *Biodiversity and Conservation*, 9(7): 901–918.

Zimmerman, M. E. (2003) The black market for wildlife: combating transnational organized crime in the illegal wildlife trade, *Vanderbilt Journal of Transnational Law*, 36: 1657.

5
Crime and the Commodification of Carbon

Reece Walters and Peter Martin

Introduction

> Carbon will be the world's biggest market. Barclays was the first UK bank to set up a dedicated carbon trading desk to help clients, and Barclays Capital is the most active player in the emissions trading market having traded 300 million tonnes as at February 2007. (Barclays, 2007: 1)

As European and American countries experience economic recession, the global emission of carbon dioxide has reached an all-time high. During a period of reduced productivity, carbon emissions have reached an unprecedented level of 34 billion tonnes (37.5 billion tons) per annum (Olivier et al., 2012). Such figures are likely to be substantially higher, with what Halsey refers to as 'the dark figure of carbon emissions' (Halsey, 2012: 169). The future picture is predicted to worsen. In March 2012, the Organisation for Economic Co-operation and Development (OECD) warned that carbon emissions were expected to rise by 70 per cent in the next four decades, increasing global temperatures by up to 6 degrees (Higgins, Short and South, 2012). This acceleration of greenhouse gas emissions is devastating news for the world's poorest peoples, where an estimated 375 million people are likely to be affected by climate-change driven humanitarian disasters in the next five years (La Chimia, 2012). At our current rate, greenhouse gas emissions are predicted to increase global temperatures by a catastrophic four to six degrees by the turn of the century (World Bank, 2012). Yet the perils of climate changing emissions are not a far-distant danger. The rapid increase of worldwide carbon emissions continues to compromise environmental sustainability whilst contributing to premature death. The link between air pollution and premature death has been widely established. It is estimated that contaminated air causes almost two million people worldwide to experience chronic respiratory malfunctions, bronchial infection, and major organ disease, resulting in ill health and 'deaths brought forward' (World Health Organization [WHO], 2011). The expansion of environmental law and regulatory arrangements in the past decade has been significant. So have various endeavours inspired by the Kyoto Protocol – namely, carbon taxes and sinks, and renewable or 'cleaner' energies.

That said, pollution or 'dirty air' has become a commodity for global trade. Section 17 of the Kyoto Protocol established mechanisms for the trade in carbon emissions; it stated that 'carbon is now tracked and traced like any other commodity. It is known as the "carbon market"'(UN Framework Convention on Climate Change, 1997). Three separate 'instruments' are used to reduce pollution: emissions trading (buying and selling emission credits); joint implementation; and the clean development mechanism investment, projects in developing countries to offset and reduce emissions (Corbera and Brown, 2010).

Numerous international attempts to regulate and enforce air emissions through permits and partnership models have proven unsatisfactory, ineffective, and counterproductive (Walters, 2010). As a result, the 'market' has become the lynchpin for corporate compliance and international monitoring. The answers to reducing global carbon emissions have been presented in recent years within discourses of trade that, as the opening quotation from Barclays asserts and incentivises, produce unprecedented profits. As a result, the world's biggest polluters, transnational corporations, have increasingly ventured into the largely unregulated voluntary carbon credit market to offset their emissions and or give their customers the opportunity to be 'carbon neutral'. The voluntary market has seen a proliferation of carbon brokers that offer tailored offset carbon products according to need and taste. This chapter examines the political economy of carbon and the ways in which trading schemes have provided new opportunities for criminal enterprises.

International developments in carbon offsetting and trading

Influenced by doctrines of green economics, the Kyoto Protocol formally accepted the use of a system of carbon credits to assist those nations unable to meet Kyoto's carbon emission goals of reducing greenhouse gas (GHG) emissions by 5.2 per cent from 1990 levels. This process of certification gives states and corporations legal permission to release emissions into the atmosphere, with one carbon credit equalling one tonne of carbon dioxide. Carbon markets aim to internalise the costs of GHG pollution within firms. As such, they provide a price signal to firms that encourage minimisation of GHG emissions or a displacement of emission savings through offsets, that is, purchasing emission savings in other firms. Markets can only occur when objects are *commensurable* and so exchange values are known and trusted. GHG instruments are remarkable to the extent that they represent no value as a tangible commodity but instead operate as a *permit* or more accurately, as a means of settling a *liability* created through GHG emissions (Mackenzie, 2009: 448). In creating this synthetic commodity, a multiplicity of projects and technologies in action must be made *commensurate*. Given this process, plus the intangibility of the instrument *and* the political basis for its value, issues of compliance, regulation, and the potential for fraud are significant concerns (Drew and Drew, 2010).

The Kyoto Protocol established upper limits or 'emission caps' for all 170 signatory countries. Large polluting nations, such as China and the United States, however, have refused to agree to mandatory caps, while the United Kingdom is widely reported to exceed its projected carbon emission target (Harvey, 2011). Moreover, polluting countries can participate in reforestation initiatives or the creation of 'carbon sinks' as a contribution to reducing emissions. The process is a trade-oriented form of control based on 'supply and demand' (Labatt and White, 2009). It was intended that rapidly developing and high-polluting countries such as China and India would need support, while other industrialised countries would need incentives in a global effort to reduce greenhouse gases. Countries now purchase carbon credits up to their maximum emission cap. Should an annual emission allowance not be met, then credits may be sold on the international market in what has become known as the global carbon trading industry. The number of consultants and traders offering advice in 'carbon finance', 'carbon accounting' and 'carbon investment' has increased substantially in the past three years (The European Business Review, 2010).

In the Kyoto system, many offset credits are sourced from projects in developing countries through the Clean Development Mechanism (CDM). This dependence of emitters on carbon reduction or avoidance in the poorer and developing world has been heavily criticised as a form of carbon 'colonisation' that provides opportunities for carbon fraud and corruption (Bachram, 2004). In the European Union (EU) Emissions

Trading Scheme (ETS), a certain proportion of credits (6 per cent in Phase Three) from the CDM can contribute to emission liabilities of entities. Significant offset credit projects are, however, also developed in high-emitting countries such as the United States and Australia, and both domestic and international credits serve the increasing voluntary market. In general, all offset credits can be purchased in the voluntary markets, but only specifically accredited credits, such as those regulated by the CDM or other national or international accreditations systems, such as the Australian Carbon Farming Initiative (CFI), can be traded in the compliance markets (Kyoto, European ETS, and proposed markets such as the Australian and Californian ETS).

In the voluntary markets, purchases are commonly from entities that wish to demonstrate corporate responsibility and attract business by growing their image as 'environmentally friendly' and enhancing branding (Peters-Stanley et al., 2011). Key buyers are major airlines and financial institutions (Hug and Ahammed, 2011: 13) but there are myriad smaller buyers from event organisers, community groups, and individuals. For most purchasers, offset credits are purchased and then 'retired' – removed from the carbon registers. Entities may also purchase carbon credits for investment purposes – for example, betting on future higher carbon prices and eventual resale, or as a pre-compliance measure to enable lower-cost compliance by buying eligible credits early. These developments in the international trading and offsetting of carbon have occurred apace in the past six years, without due consideration for regulatory and enforcement practices. As a result, and perhaps unsurprisingly, international headlines such as 'Interpol warns of carbon fraud' (Packham, 2009) and 'Fraud mastermind cheated taxpayers of 39 million pounds in just 69 days' (Allen, 2012) have provided wake-up calls for governments and industries focussed on the benefits of trade rather than on the opportunities of criminal entities. The emergence of carbon fraud rhetoric has provided necessary political reflection on the perils and prospects of the market's role in solving the market's problems.

Frauds and risks in carbon markets

Fraud is notoriously difficult to define at law (Beull, 2011). There are various questions of authenticity, legitimacy, acceptability, and imposture that are relevant to legal notions of misrepresentation and dishonesty that underpin regulatory authorities' decisions to pursue prosecution. Such decisions have been exacerbated with carbon. The carbon market poses complex conceptual questions about the nature of carbon as a tradeable item. If carbon is a commodity, then it should be subject to VAT like all other tradeable properties. But is it personal or real property? Most legal jurisdictions have not provided case law decisions on carbon; it is generally referred to as a 'property-like right'. This lack of legal definition and clarity has ramifications for both regulation and cross-national trading.

The media have widely reported the risk of carbon fraud in both the allowance markets (for example, the European ETS) and in the offset credit market. Lohmann (2010) has stated that 'uncovering carbon market scandals is now a minor journalistic industry'. Large accounting firms are informing their clients and releasing scoping documents such as Deloitte's: 'Carbon Credit Fraud – The White Collar Crime of the Future' and more recently 'Carbon Credit Fraud – an update' (Deloitte, 2009). Moreover, tradeable carbon allowances rely upon polluting industries to honestly disclose emissions. Emissions reporting has regularly been identified by accountants as a carbon fraud risk (see Deloitte 2009: 3; Lindquist and Goldberg, 2010: 63). Regulatory and policing authorities in the European Union do not have the resources to conduct the necessary proactive inspections to ensure compliance. In addition, fraud potentially exists with exaggerated or falsified estimations of carbon benefits from the proposed projects. Barr (2011: 335), drawing on earlier work by Ross (2001), also makes the point that powerful state actors could find it financially rewarding to overreport emissions for short periods of time.

As is often the case with the identification of new forms of green crime, resistance and protest groups provide the initial impetus for subsequent criminal justice concerns (Walters, 2012). The most sustained critiques on carbon credit fraud risk have come from nongovernmental organisations (NGOs) such as Transparency International (2011), Global Witness in their report 'Forest Carbon Cash and Crime' (2011), and Greenpeace's 'Carbon Scam' (Densham et al., 2009). The Corner House has maintained a sustained critique of carbon trading over the last decade (The Corner House, 2001; Gilbertson and Reyes, 2009) with the same mention of criminal fraud (ibid., 73) and corruption (ibid., 63). These reports challenge the authenticity of carbon credit schemes, particularly in REDD (UN Reducing Emissions from Deforestation and Forest Degradation) subnational projects, and warn of the risks of criminal engagement in these schemes. Most of these reports identify risks rather than actual cases, although a number of anecdotes are given, and in the case of 'Carbon Scam', Greenpeace analyses in detail the Noel Kempff Climate Action Project in Bolivia and identifies major shortcomings in the net emissions savings and 'avoided deforestation' (Densham et al., 2009: 9–13).

There has been significant fraud in the EU ETS compliance market through value-added tax (VAT) scams (carousel fraud) and Internet 'phishing'. In addition, there have been isolated instances of fraud suspected in the 'broking' of carbon offset credits in the voluntary market, and investment scammers have moved into the voluntary carbon market (Walters and Martin, 2012).

Carousel fraud

In September 2012, seven individuals were prosecuted in the United Kingdom, charged with VAT fraud from trading in EU carbon allowances. They were involved in complex 'carousel trades', which utilised the VAT-free export of securities within the EU and sold them on with VAT included in the sale, without paying VAT to the government. This instance of VAT 'skimming' netted £38 million in 69 days, and reports further suggest that this type of fraud has cost the EU countries around €3 billion (Allan, 2012). Four of the defendants were acquitted, but three were found guilty and sentenced up to 15 years imprisonment (Harries, 2013).

This case resulted in the UK government making carbon transactions exempt from VAT (Harries, 2012). Internet fraudsters have also sent e-mails to companies using the EU ETS registries that directed them to websites where they were asked to enter their identification number and passwords. These were used to sell on emission allowances. This scam, known as 'phishing', was reported to have costs the companies millions of euros.

Offset schemes and the exploitation of indigenous peoples

The Kyoto Protocol established 'offsetting carbon emissions' through the financing of emission reduction projects or 'carbon sinks' in other parts of the globe. The Kyoto Protocol is a baseline and trade market where richer countries are expected to meet their emission targets by three mechanisms: first, by purchasing credits from other developed countries who have reduced emissions; second, by the Joint Implementation (JI) mechanism, which allows purchasing project-based offset credits from other countries with binding targets; and third, through the Clean Development Mechanism (CDM), which allows the purchasing of project offset credits from developing countries that have no binding emission targets (Hepburn, 2007: 379). Credits from offset projects are also utilised in the voluntary market. Emitting entities can purchase credits from carbon brokers to partially or fully offset their GHG emissions. The voluntary market is largely unregulated by the state and, whilst being regarded initially as the 'wild West' of carbon markets, it has seen increasing self-regulation and greater standards of accreditation and certification (Hamilton et al., 2008: 53). That said, accusations of bribery have been made in Liberia in large carbon credits deals (Global Witness, 2011). In Papua New Guinea (PNG) and Peru, accusations have been made that private developers and NGOs have been swarming the tropical forests to encourage indigenous leaders to sign away their rights to the forest. In these instances, it has been reported that Peruvian indigenous peoples, for example, have been pressured to sign agreements that they cannot read in acts described as 'carbon piracy' (Vidal, 2011). Moreover, serious accusations were made of PNG officials producing fake carbon credit certificates as a prop for explaining carbon credit deals to local leaders (Wilkinson and Cubby, 2009).

Another Australian-based carbon broker, Shift2Neutral, has been accused of distributing fake carbon offset credits and providing no evidence of having successfully negotiated the project with local indigenous peoples (Cubby, 2011). The Carbon News Forum (CNF) reported:

> Alarm bells about Shift2Neutral have recently been rung by the Tribal Coalition of Mindanao, who in late November 2010 report that a 17-month-old $500 million Tricom Caraga Memorandum of Understanding between Shift2Neutral and indigenous tribes of Caraga has been dissolved. The tribal people believe that they have been conned. This follows on from reports dated 6 October 2010 that a recently signed 'Shift2Neutral agreement in Congo was illegal'. (Carbon News Forum, 2011)

Accusations of illegality in offset projects have mainly been in tropical forest contexts in developing countries. Here, the structural conditions for fraud and corruption mean that projects could be more risky and that the establishment of 'carbon credit' forests could occur through deception or bribery. Such conditions include geographical remoteness, weak governance and official corruption, low levels of economic development, and uncertainty over land ownership (Contreras-Hermosilla, 2002). The uncertainty of land tenure provides particularly lucrative opportunities for criminal entities. It is estimated that throughout the world over two billion customary land owners are not recognised in national laws (Wiley, 2008). Such conditions, which often occur in countries exploited by illegal loggers, exist in those nations where the newly accredited United Nations Reducing Emissions from Deforestation and Forest Degradation (REDD+) programme will occur. This programme produces carbon credits for forest sink conservation and development, and expects to supply funds between US$17 and $33 billion every year, much of which will be destined for forest-rich developing countries. The programme is not just project-based, but intends to compensate governments, communities, companies, and individuals in developing countries who undertake to reduce emissions loss from forests. The huge sums of money involved and structural conditions in these countries have NGOs greatly concerned with potential fraud, corruption, and bribery (Global Witness, 2011).

Verification fraud

There are also a number of studies (Barr, 2011; Brown, 2010; Drew and Drew, 2010) that point to the vulnerability of verification and validation processes in the CDM process, the largest offset scheme. Attention to deficits in the CDM process was heightened in 2006 when spot checks by UN inspectors found significant irregularities in work by three prominent verifiers (Schneider, 2007: 24). In 2009, UN inspectors suspended the largest verifier, SGS United Kingdom, because of poor-quality documentation and lack of adequate qualifications of their staff. The process of verification is integral to safeguarding against fraud. However, the offset verification industry is very price competitive, with revenue per project declining and verifiers highly dependent on the project developers (Brown, 2010). As project developers try to cut costs, they search for the verifiers with the lowest fees, and the issue of verification quality is not being considered as an important price factor. Further, oversight of verifiers by CDM is limited by insufficient resources (Schneider, 2007). Concerns have also been expressed regarding firms providing consulting advice to project developers and acting as verifiers for the project (Bachram, 2004: 5).

Human rights abuse and carbon offsets

The absence of compliance trading systems in many countries and the growing social awareness of the impact of climate change have created a growing market for the *voluntary* carbon offset market. The primary market here is for voluntary purchases from entities that wish to demonstrate corporate responsibility and attract business by growing their image as 'environmentally friendly' and enhancing their branding.

To achieve this, offset credits are purchased and then 'retired' – removed from the carbon registers. Entities may also purchase carbon credits for investment purposes – for example, betting on future higher carbon prices and eventual resale, or as a pre-compliance measure to enable lower-cost compliance by buying eligible credits early.

There is significant differentiation of carbon offset products, and price is very much related to perceived quality of the projects such as the incorporation of sustainable development criteria, benefits for local people and their environment, and projects that are relatively free from fraud risk. This has led to high-priced specialist 'boutique' carbon markets, driven by buyers from developed countries in times of excess credits and / or buyers want to increase their public image through projects that promote strong environmental and social benefits (Bumpus, 2011: 623). A part of ensuring quality is purchasing from registries that ensure transparency and reduce the risk of double counting of credits. Further, certification by reputable verification standards, such as the Voluntary Gold Standard (VGS), promotes a positive image to the purchaser and reduces risk of funding mismanagement or fraudulent projects.

Of considerable concern is the creation of carbon offset projects in regions afflicted with conflict and human rights abuses. For example, the Bajo Aguan Valley in Honduras witnessed the murder of 23 farmers between January 2010 and March 2011 over rights to land (Euractiv, 2011). More than 3,500 peasant farmers have been petitioning for their right to land for agricultural purposes since the coup d'état on 28 July 2009. Independent international human rights inspectors observe that government officials in collaboration with private security firms have systematically oppressed the rights of local and indigenous peoples and have orchestrated executions to prevent further protest, stating:

> The government has converted the area of these agrarian conflicts in Bajo Aguan into a war zone: low-flying military helicopters and planes, armed commandos passing menacingly through defenceless villages during the days after the coup; and the peasants of the region's organized movement suffer kidnappings, torture and murders. Human rights violations are growing as the peasants' claims increase. (International Federation of Human Rights, 2011)

Peasant farmers were attempting to reclaim land from Groupo Dinant, a large corporation that had converted disputed land into palm oil plantations that the Honduran government registered as a carbon offsets for European polluters (Nelson, 2011). This issue has raised ethical and moral issues about carbon offsetting in countries with reported human rights violations and about the extent to which the commercialisation of carbon serves to exacerbate the plight of those abused.

Investment and carbon price rebate scams

The *modus operandi* in these scams is 'cold calling', either through e-mail or by telephone. In 2011, the Australian Transaction Reports and Analysis Centre (AUSTRAC) reported a fake carbon credits investment scheme that cost investors AUD 3.5 million (AUSTRAC, 2011: 28–29). The victims of the scheme were mostly small business investors and self-funded retirees who were interested in ethical investments. Telemarketers made unsolicited calls and discussed environmental concerns with the aim of drawing them in to the investment scam. If interested, investors would be contacted by a representative of a Japanese-based investment scam business, who would offer them the opportunity to invest in overseas carbon credits. Those who took up the offer would be asked to transfer money to accounts in Taiwan and China. A professional website had been constructed to allow victims to view their investment certificates. AUSTRAC's attention to the scam was initiated by two very large follow-up transfers. Victims subsequently reported that they had no access to their certificates and they could not be liquidated.

Similar investment scams have been found in countries with high visibility compliance and voluntary carbon emissions trading schemes. The high public awareness of the monetisation of carbon seems to lead to opportunities for scammers to attract 'investment' money. The Financial Services Authority (FSA) in the United Kingdom has recently alerted the public about a host of fraudulent carbon investment schemes that are offered to investors by salespeople, e-mails, telephone, post, or even by 'word of mouth' (Financial Services Authority, 2012).

Concluding comments

As South and Brisman (2013: 99) remind us, green criminology is the 'pursuit of social justice and human rights'. Such questions inevitably involve an examination and intersection of concepts of harm, power, and justice. This chapter identifies the ways in power is mobilised to justify a market model of capitalism, with unjust and harmful consequences for the environment and the world's most vulnerable peoples.

Carbon markets are an extension of neoliberal governance and the commodification of nature (see Thornes and Randalls, 2007; Pearse, 2011; Bailey et al., 2011; Paton and Bryant, 2012). For some, carbon fraud is symptomatic of wider systemic properties of carbon marketisation that are inherently corruptible (Lohmann, 2010) and fraudulent (Bachram, 2004), and this connects with the concerns of NGOs. Others, however, argue that 'marketisation' of carbon is a political project itself and can be shaped by social concern for its negative consequences (MacKenzie, 2009).

Fraud risk is a major concern in market development. There are various tensions in the carbon markets that may further determine how risk is avoided, minimised, or incorporated into the market. Because of the difficult and tenuous nature of carbon commodification, transaction costs are quite high and the calls for increased regulation and oversight to deal with issues such as fraud will increase these costs and could affect the supply of credits in situations of low carbon prices in the compliance markets. On the other hand, degradation in quality of offsets through inadequate oversight could lead to rejection by compliance markets and increasing risk for voluntary purchasers.

Carbon 'broking' fraud is a high-risk area. There is currently only weak regulation through traditional consumer laws. United Nations oversight is stretched and verification is dominated by a few large companies in a highly competitive environment, and some verifiers are also involved in consulting for project developers. UN spot checks and reviewers have found many verification reports to be unsatisfactory in terms of project additionality, monitoring and estimation of carbon saved, and local consultation. The experience of illegal forestry indicates that a higher risk of criminality exists in those countries with projects that are geographically remote and/or have low levels of economic development, weak governance and state institutions, and unclear land tenure systems. Misrepresentation and fraudulent reporting of emissions by liable entities is a significant risk, although there are few reports of this in the established EU ETS.

Carbon is a unique product for marketisation. Customers who invest in carbon offsets projects rely heavily on brokers for advice. Unlike most 'commodities', carbon is a pollution without a designated origin that cannot be assessed for market performance and due diligence. The complex carousel frauds in Europe have resulted in governments providing VAT exemptions from carbon transactions. Why should polluting industries be given tax exemptions because of regulation failures? The answer points to the centrality in the market model favoured by EU countries. Moreover, this chapter concludes that examples of fake offsets, fabrication of carbon certificates, bribery of government officials, and the exploitation of and violence towards indigenous and poor peoples in developing countries point to a capitalist enterprise having devastating social, political, and environmental impacts. So, concerned by the unethical, illegal, and unregulated activity of emissions trading and carbon markets; untrustworthy actions of carbon brokers; and the unreliability of carbon emission trading, the Financial Services Authority in the United Kingdom has recently released a stark public warning:

> Carbon credits can be sold and traded legitimately and there are many reputable firms operating in the sector. However, we are concerned that an increasing number of firms are using dubious, high-pressure sales tactics and targeting vulnerable customers. We do not regulate carbon credits as a product in the same way as shares or units. This means a firm promoting or selling them does necessarily have to be authorised by us. (Financial Services Authority, 2012, cited in Walters and Martin, 2012)

This admission, by the key share market regulator in Britain, identifies how the regulators of carbon trading have little control over illegal activities.

References

Allen, E. (2012) 'Fraud mastermind cheated taxpayers of 39 million pounds in just 69 days', *The Daily Mail.* Available at http://www.dailymail.co.uk/news/article-2161520/Fraud-mastermind-cheated-taxpayer-39million-just-69-DAYS-spent-money-1million-home-Rolls-Royce-jailed-15-years.html. Accessed 10 December 2012.

AUSTRAC (2011) *AUSTRAC Typologies and Case Studies Report 2011.* Canberra: Australian Government.

Bachram, H. (2004) 'Climate fraud and carbon colonialism: the new trade in greenhouse gases', *Capitalism Nature Socialism,* 15: 1–16.

Barclays (2007) *Products and Services for Business.* Available at www.personal.barclays.co.uk/BRC1/jsp/brccontrol?task=articleFW. Accessed 2 March 2013.

Barr, C. (2011) 'Governance risks for REDD+', 6.3 in Transparency International. *Global Corruption Report: International Climate Change.* Earthscan: London.

Brown, M. (2010) 'Limiting corrupt incentives in a global REDD regime'. *Ecology Law Quarterly,* 37: 237–267.

Bumpus, A. (2011) 'The matter of carbon: understanding the materiality of CO_2 in carbon offsets', *Antipode,* 43: 612–638.

Carbon News Forum (2011) *Summary of Brett Goldsworthy Shift2Neutral con man.* Available at http://shift2nothing.com/2011/04/summary-of-brett-goldsworthy-shift2neutral-con-men/. Accessed 14 January 2013.

Contreras-Hermosilla, A. (2002) *Law Compliance in the Forestry Sector: An Overview.* WBI Working Papers. Washington, DC: The World Bank.

Corbera, E. and Brown, K. (2010) 'Offsetting benefits? Analyzing access to forest carbon', *Environment and Planning*, 42: 1739–1761.

Cubby, B. (2011) 'Revealed. Scandal of carbon credit firm', *The Sydney Morning Herald*. Available at http://www.smh.com.au/environment/revealed-scandal-of-carbon-credit-firm-20110407–1d6a4.html. Accessed 15 January 2013.

Deloitte (2009) 'Carbon Credit Fraud – The White Collar Crime of the Future'. Available at http://www.deloitte.com/view/en_AU/au/insights/ca42b7c6a-b0e1310VgnVCM1000001a56f00aRCRD.htm. Accessed 3 February 2013.

Densham, A., Czebiniak, R., Kessler, D. and Skar, R. (2009) *Carbon Scam: Noel Kempff Climate ACTION projects and the Push for Sub-national Forest Offsets*. The Netherlands: Greenpeace International. Available at http://www.greenpeace.org/france/PageFiles/266591/carbonscam.pdf. Accessed 29 November 2012.

Drew, J. M. and Drew, M. E. (2010) 'Establishing additionality: fraud vulnerabilities in the clean development mechanism', *Accounting Research Journal*, 23(3): 243–253.

Euractiv (2011) Carbon credits tarnished by human rights 'disgrace'. http://www.euractiv.com/climate-environment/carbon-credits-tarnished-human-r-news-508068 accessed 5 April 2013.

The European Business Review (2010) *Noble Carbon Credits: Emissions in Trading is a Comparatively New Market.* Available at www.europeanbusinessreview.co/?p=331. Accessed 30 November 2012.

Financial Services Authority (2012) 'Carbon Credit Trading'. Available at http://www.fsa.gov.uk/consumerinformation/scamsandswindles/investment_scams/carbon_credit. Accessed 2 March 2013.

Global Witness (2011) *Forest Carbon, Cash and Crime: The Risk of Criminal Engagement in REDD+*, September.

Halsey, M. (2012) 'Defining pollution down: forestry, climate change and the dark figure of carbon emissions', in S. Farrall, T. Ahmed and D. French (eds) *Criminological and Legal Consequences of Climate Change*. Oxford: Hart Publishing.

Hamilton, J., Sjardin, M., Marcello, T. and Xu, G. (2008) *Forging a Frontier: State of the Voluntary Carbon Markets 2008.* A report by Ecosystem Marketplace and Bollomberg New Energy Finance.

Harries, M. (2013) 'Fraud and financial crime'. Available at http://www.carmelite-chambers.co.uk/members/mark-harries/expertise/fraud-and-financial-crime-mark-harries. Accessed 2 March 2013.

Harvey, F. (2011) 'Coalition on course to miss carbon emissions targets, says CBI', *The Guardian.* http://www.guardian.co.uk/environment/2011/jun/08/coalition-miss-emissions-targets-cbi. Access 5 April 2013.

Hepburn, C. (2007) 'Carbon trading: a review of the Kyoto mechanisms', *Annual Review of Environmental Resources*, 32: 375–393.

Higgins, P., Short, D. and South, N. (2012) 'Protecting the planet after Rio – the need for a crime of ecocide', *Criminal Justice Matters*, 90: 4–5.

International Federation of Human Rights (2011) Honduras: Human Rights in Bajo Aguan, http://www.fidh.org/IMG/pdf/honduras573ang.pdf, Accessed 5 April 2013.

Labatt, S. and White, R. (2009) *Carbon Finance: The Financial Implications of Climate Change*. Toronto: Wiley.

La Chimia, A. (2012) 'Climate change and development aid funds', in S. Farrall, T. Ahmed and D. French (eds) *Criminological and Legal Consequences of Climate Change*. Oxford: Hart Publishing.

Lindquist, C. S. and Goldberg, S. R. (2010) Cap-and-trade: accounting fraud and other problems, *The Journal of Corporate Accounting and Finance*, 21(4): 61–64.

Lohmann, L. (2010) 'Regulation as corruption in the carbon offset markets', in S. Bohm and S. Dabhi (eds) *Upsetting the Offset: The Political Economy of Carbon Markets*. Zed Books: London.

MacKenzie, D. (2009) 'Making things the same: gases, emission rights and the politics of carbon markets', *Accounting, Organizations and Society*, 34: 440–455.

Nelson, A. (2011) EU carbon credits scheme tarnished by alleged murders in Honduras – Calls for Honduran credits to be thrown out of ETS following alleged killings, while consultation process goes under review, *The Guardian*. http://www.guardian.co.uk/environment/2011/oct/03/eu-carbon-credits-murders-honduras. Accessed 3 October 2011.

Packham, B. (2009) 'Interpol warns of Carbon fraud', *Adelaide Now*, http://www.adelaidenow.com.au/news/interpol-warns-of-carbon-fraud/story-e6freo8c-1225733540371. Accessed 2 March 2013.

Peters-Stanley, P., Hamilton, K., Marcello, T. and Sjardin, M. (2011) *Back to the Future: State of the Voluntary Carbon Markets 2011*. A report by Ecosystem Marketplace and Bollomberg New Energy Finance.

Ross, M. (2001) *Timber Booms and Institutional Breakdowns in South East Asia*. Cambridge University Press: Cambridge.

Schneider, L. (2007) *Is the CDM Fulfilling its Environmental and Sustainable Development Objectives?* An evaluation of the CDM and options for improvement. Berlin: WWF, Institute for Applied Ecology.

South, N. and Brisman, A. (2013) 'Critical green criminology. Environmental rights and crimes of exploitation', in S. Winlow and R. Atkinson (eds) *New Directions in Crime and Deviance*, London: Routledge, pp. 99–111.

Transparency International (2011) *Global Corruption Report: International Climate Change*. Earthscan: London.

United Nations Framework Convention on Climate Change (1997) *Kyoto Protocol*. Available at http://unfccc.int/kyoto_protocol/items/2830.php. Accessed 3 March 2013.

Vidal, J. (2011) 'Peruvian Amazon could become global centre of "carbon piracy": report', *The Guardian*. Available at http://www.guardian.co.uk/environment/2011/nov/30/peruvian-amazon-carbon-piracy?INTCMP=ILCNETTXT3487. Accessed 30 November 2011.

Walters, R. (2010) *Crime is in the Air. The Politics and Regulation of Pollution in the UK*. London: CCJS.

Walters, R. (2012) 'Dirty collar crime and the environment', *Proceedings of the Crime, Justice and Social Democracy: An International Conference*. Brisbane: QUT, pp. 103–112.

Walters, R. and Martin, P. (2012) *Risks of Carbon Fraud*. Brisbane: Centre for Crime and Justice Research.

Wilkinson, M. and Cubby, B. (2009) 'Carbon scandal snares Australian', *The Age*, Available at http://www.theage.com.au/business/carbon-scandal-snares-australian-20090903-f9yy.html. Accessed 3 March 2013.

World Bank (2011) *State and Trends of the Carbon Market 2011*. Washington, DC: The World Bank.

World Bank (2012) *Turn Down the Heat. Why a 4-degree Warmer Planet Must Be Avoided*. Available at http://climatechange.worldbank.org/sites/default/files/Turn_Down_the_heat_Why_a_4_degree_centrigrade_warmer_world_must_be_avoided.pdf. Accessed 3 March 2013.

World Health Organization (2011) 'Air quality and health', Factsheet No. 313, September 2011. Available at http://www.who.int/mediacentre/factsheets/fs313/en/index.html. Accessed 3 March 2013.

6

The Local Context of Transnational Wildlife Trafficking: The Heathrow Animal Reception Centre

Tanya Wyatt

Introduction

Wildlife trafficking most likely conjures up images of a faraway seedy exotic street market full of cages of diverse wildlife, bound for stewpots or a collector on the other side of the globe. This chapter teaches us that 'the other side of the globe' is often as local as the nearby airport. Almost daily, illegally trafficked wildlife is transported through, or stored within, the surrounding industrial neighbourhood of London's Heathrow Airport. Confiscated animals that are protected under the Convention on the International Trade in Endangered Species of Wild Fauna and Flora (CITES) are seized at Heathrow and sometimes around the country and are brought to London to be housed at the Heathrow Animal Reception Centre, known as HARC. This chapter details the unique regulatory position and oversight given to HARC and examines the official documentation of seizures by the CITES enforcement authority; it also explores issues of harm and justice that arise from wildlife trafficking. Data were obtained by speaking with HARC staff and touring the facility, as well as by interviewing a CITES enforcement officer. The information gathered provides context as to how the transnational green crime of wildlife trafficking affects the local situation. Further evidence of the local impact is evident through a closer examination of two cases of animal confiscation that took place at HARC. These cases bring to life the realities of the injustices and harms that are inherent in the illegal wildlife trade, which are often unstated or unexplored.

The Heathrow Animal Reception Centre (HARC)

The legal trade in wildlife is annually a multibillion-dollar industry that involves hundreds of millions of plants and animals being shipped and transported around the globe (CITES, 2012). Transportation of live animals over short and long distances inevitably raises concerns about harm in relation to animal welfare. In a green criminological framework, this leads, in turn, to issues of species justice (White, 2011), where individual animal suffering is worthy of further exploration. The sheer volume of this industry requires specialised infrastructure and policy apparatus to regulate and scrutinise the transactions that are taking place. This means that certain authorised airports and shipping ports have dedicated personnel and facilities to inspect and store containers of legal wildlife.

In the United Kingdom, this task predominantly falls to the Heathrow Animal Reception Centre, where most of the wildlife entering the United Kingdom arrives, as well as a majority of farm animals and pets. Moreover, since HARC has qualified staff and facilities to house various animals, councils and charities like the Royal Society for the Prevention of Cruelty to Animals (RSPCA) reach out to HARC for help in caring for exotic wildlife that is sometimes seized in other contexts (personal communication HARC staff, 20 April 2011). In terms of the legal trade, though, HARC staff are tasked with verifying the paperwork that indicates that the live animals are legal and healthy for all shipments. Additionally, staff inspect the shipments for smuggled wildlife, stowaways, and animal welfare violations (personal communication HARC staff, 20 April 2011). In essence then, assessing harm is a priority for staff.

Background and overview

HARC sits in a unique regulatory and governance position. It is part of the local authority of the City of London's Environmental Services Directorate, Veterinary Sector (City of London, 2012). This local authority is responsible for the 'Square Mile' of central London, yet historically the expertise and success of the local authority in regard to animal welfare and containing rabies meant that it took on this remit for the Greater London area (City of London, 2012). This includes Heathrow Airport and the business district that surrounds it, where HARC is now located. The City of London then, through HARC, has the statutory responsibility of the Animal Health Act 1981 (UK) and the Welfare of Animals During Transport Order 1997 (UK). These acts are part of the legislation that complies with the European Union's standards for animal health and welfare established in Council Regulation (EC) No 1/2005 (City of London, 2012). The Animal Health Act 2002 gives HARC the authority to treat, quarantine, and slaughter diseased animals as well as check that they do not suffer during transport. The Welfare of Animals During Transport Order 1997 (UK) similarly ensures the welfare of animals while being shipped and also requires compliance with welfare standards and regulations regarding transport set by CITES for the species listed in their appendices. CITES (1979) advises carriers and transporters to monitor the conditions and health of all live animals during transport. This means checking for sick animals, and checking that ventilation is sufficient and that there are not extreme temperature fluctuations (CITES, 1979). CITES (1979) also advises about the materials of containers that should be used for transporting different species and gives guidelines as to the availability of food and water. The containers must be clearly marked that they contain live animals, with instructions on not tipping or overturning the crate highly visible (CITES, 1979).

Connected closely to the Animal Health Act 1981 is the Live Animals 'Balai' Directive 1992, which HARC also enforces. This directive provides guidelines for the trade in those species that are not covered elsewhere in EU legislation (Department of Environment, Food and Rural Affairs [DEFRA], 2011). This pertains to member states and to imports from third countries (DEFRA, 2011). It essentially puts into place health and disease monitoring of primates, ungulates not in the meat industries, birds, dogs, cats, rabbits, bees, other mammals, and other animals susceptible to rabies (DEFRA, 2011). As mentioned, this covers animals in trade and in particular animals held for display, education, conservation or research programmes, and laboratory animals (DEFRA, 2011). When importing wildlife to the United Kingdom, further documentation is needed that the importer has received permission from the local authority to own and house wildlife that is listed in the Dangerous Wild Animals Act 1976. This includes most primates, big cats, nondomesticated canines, elephants, and a range of other species, all of which HARC would verify documentation for.

Official inspection of the documentation, the welfare, and the health of live animals requires numerous expert staff from multiple agencies, time, and resources. The reception centre itself is staffed 24 hours a day, 365 days of the year, since live animals are continually arriving into the United Kingdom, and there are always animals housed awaiting inspection or clearance and therefore requiring care (personal communication HARC staff, 20 April 2011). This is particularly true of birds that enter the country, which must be quarantined for a certain period of time to ensure they are not carrying diseases; as there is no commercial quarantine, all birds are housed and cared for at HARC (personal communication HARC staff, 20 April 2011). Care is given to wild animals, farm animals, and pets, although this chapter focuses on the wild animals.

HARC has around 20 staff members, most of who are animal care workers (personal communication HARC staff, 20 April 2011). These staff coordinate with staff of DEFRA, who issue the Border Crossing Certificates for wildlife entering the United Kingdom from outside the European Union (City of London, 2012). If needed, further veterinarian assistance can also be drawn upon from the City of London (personal communication HARC staff, 20 April 2011). HARC staff conduct the inspections of live animals being transported; the manager and deputy managers inspect all of the CITES and venomous shipments (personal communication HARC staff, 20 April 2011). Additionally, the HARC staff coordinate with the CITES enforcement team who are law enforcement officers of the UK Border Agency. The CITES enforcement team is stationed full-time at Heathrow Airport, searching for smuggled wildlife and wildlife products and checking CITES documentation of CITES shipments (personal communication CITES enforcement officer, 26 June 2011). They are responsible for the CITES animals (and plants) traded, whereas HARC is responsible for assessing if there is harm to animal health and welfare (personal communication HARC staff, 20 April 2011), which, as will be made evident in the next section, is a daunting task.

The wildlife: facts and figures

In 2010, 185 million animals came through HARC – this again is wildlife, farm animals, and pets, which includes large-quantity shipments like ants and bee larvae (personal communication HARC staff, 20 April 2011). This consisted of approximately 10,000 cats and dogs, 900 'other' mammals, 2000 birds, 150,000 bird chicks, 300 horses, 300,000 reptiles and amphibians, 34 million fish, and several million invertebrates (personal communication HARC staff, 20 April 2011). It is impossible to closely inspect each of these shipments, so HARC uses a risk-based model for inspecting approximately 10 per cent of shipments for smuggling, stowaways, and health and welfare issues (personal communication HARC staff, 20 April 2011).

Legal shipments of wildlife may be used as a way to smuggle wildlife and may inadvertently contain exotic wildlife that can be harmful to local ecosystems. HARC staff are trained in species identification so that when inspecting shipments they can determine if the species identified on the documentation is in fact the species that is in the container as well as identify the species of any stowaways (personal communication HARC staff, 20 April 2011). If found, both misdeclared species and stowaways would be confiscated and housed indefinitely at HARC until they can be found a new home at a zoo or wildlife refuge (personal communication HARC staff, 20 April 2011). For example, a sulcata tortoise, which is listed in CITES Appendix II (threatened, with trade allowed within set quotas), was documented as another kind of non-CITES tortoise and was confiscated by the United Kingdom Border Agency (UKBA) CITES enforcement team and was being housed at HARC (personal communication HARC staff, 20 April 2011). In addition to misdeclaring species within shipments, wildlife traffickers also put undeclared wildlife in with legal wildlife. For instance, a shipment within the last two years was of a consignment of fish, but also contained red-eared terrapins, not a protected species, but a turtle species popular in the pet trade that is cause for concern because it has become invasive in certain parts of the world (personal communication HARC staff, 20 April 2011).

Finally, traffickers will fill shipments of wildlife with more animals than is declared on the paperwork. UKBA seized part of a shipment of day geckos because the paperwork had indicated a specific number, but there were several times that number in the container (personal communication HARC staff, 20 April 2011). It should be noted in terms of environmental impact that, particularly in the reptile trade, most of the animals that are inspected at HARC are from captive-bred stocks rather than from the wild (personal communication HARC staff, 20 April 2011). This still means that if they belong to a species listed within the CITES appendices, they need the required import and export permits.

Shipments of wildlife are also checked for their compliance to animal welfare regulations during transport. For containers arriving at HARC from Heathrow Airport, this means that they must comply with the previously mentioned legislation, which is reflected in the established guidelines of the International Air Transport Association (IATA). Ideally, the airline will have the exporter open the container of wildlife before the airline accepts it for shipment to ensure that the proper welfare standards are being met (personal communication HARC staff, 20 April 2011). Not only is this good practice, but it is advisable in terms of responsibility and welfare of the wildlife (personal communication

HARC staff, 20 April 2011). In essence, when HARC inspects a shipment and they find a welfare violation, it is the airline that is contacted rather than the exporter. Responsibility for the harm to the wildlife and the welfare violations is difficult to prove (personal communication HARC staff, 20 April 2011). All containers must be properly labelled on the outside and there must be air holes, and not too many animals within any one container (personal communication HARC staff, 20 April 2011). For instance, in a shipment that was inspected during this research a small plastic box, which was one of multiple small containers in a much bigger crate, contained 20 small turtles. The container should have only had eight turtles so that each of them would have space to stand on the box rather than on top of each other. HARC staff, rather than initiating any formal process surrounding animal welfare in this instance, decided to speak with both the airline and the exporter to clarify shipping conditions.

Particularly in the pet trade of reptiles, the HARC staff have working relationships with the importers and exporters who they have frequent contact with. HARC staff also do outreach and training internationally with airlines in order to improve animal welfare during transport. They are in the middle of an industry that is trying to balance animal welfare and profitability by maximising their shipments (personal communication HARC staff, 20 April 2011). This at times appears to result in overloaded containers and those without the proper standards, such as air holes and labelling, but also leads to more serious smuggling tactics.

Seizures and violations

Violations of welfare standards are grounds for seizure, as is missing or fraudulent paperwork. This indicates that harm to animals is taken seriously in the United Kingdom and the European Union. Seizures have been dropping in the last few years, and 2011 started off with only one CITES violation (personal communication HARC staff, 20 April 2011). HARC staff felt that welfare during transportation was getting better and that there was less smuggling, but such incidents are extremely variable (personal communication HARC staff, 20 April 2011). Shipments are very seasonal and to date this year (2011) have been very low. To compare, between 1996 and 2000 there were 1001 seizures, 17 per cent of which were live animals (World Wide Fund for Nature [WWF], 2002). In 2007, the percentage of live animal confiscations had dropped to 8 per cent of total seizures (DEFRA, 2007).

The seizure history shows that the detected smuggling at Heathrow consists mainly of reptiles and that welfare violations are sometimes but not always a feature of the trade. In the late 1990s, the CITES enforcement team, then part of Customs, seized several large-quantity mislabelled shipments of CITES-listed reptiles. In 1997, 294 Kenyan sand boas coming from Zambia and going to the United States were documented as wild specimens, although sand boas alleged to be coming from Zambia (TRAFFIC, 2011). Several hundred geckos, chameleons, and Mantella frogs (nonpoisonous, but with markings like a poison dart frog) had no paperwork and were also seized (TRAFFIC, 2011). All of these animals were cared for by HARC staff and then given to zoos or aquariums. In 1998, 15 Horsfield's tortoises coming from Uzbekistan were confiscated because there were more specimens than the permit allowed. Additionally, a shipment of sea turtle eggs from Brunei was not collected by the owner, but would have been seized anyway as they are banned from trade (TRAFFIC, 2011). The following year, 100 spectacled caimans and multiple species of boa constrictors, anacondas, and tortoises were seized as their export and import permits had expired (TRAFFIC, 2011).

Large-quantity confiscations of reptiles continued into the 2000s. A shipment of 149 spiny-tailed lizards coming from Sudan through Spain contained 68 Appendix II listed Bell's Dabb lizards, which had been labelled as the unlisted Eyed Dabb lizard, which was the species of the rest of the lizards in the container (TRAFFIC, 2011). Then in 2001 a passenger from Russia on the way to Tunisia was carrying three suitcases with 710 Appendix II Horsfield's tortoises without CITES documentation and in violation of the IATA's welfare regulations (TRAFFIC, 2011). The person was arrested and the tortoises were taken to HARC. Ten CITES Appendix I species of crocodiles and 25 Appendix II species of pythons and monitor lizards were confiscated when Customs officials discovered that the paperwork claiming the shipped animals had been farmed in Benin had been forged. The container was inspected in transit from Nigeria to South Korea. Once again, HARC housed the animals until those that survived were given to breeding facilities. In this instance, several of the crocodiles died from internal injuries suffered from the hooks used to capture them, which were still embedded in their bodies (TRAFFIC, 2011). In late 2003, a shipment of Appendix II chameleons arriving from Benin and headed for the United States had documentation that listed them as a different chameleon species and the paperwork accounted for only 50 specimens when there were 98 (TRAFFIC, 2011). No confiscations of reptiles took place again until 2006, when 240 spectacled caimans were seized by CITES enforcement officials for having an invalid export permit (TRAFFIC, 2011). The shipment originated in Guyana and was destined for Russia. There were several seizures of reptiles in 2008 by the CITES team, which at that point was part of the UK Border Agency. First a shipment of 110 leopard tortoises coming from Zambia was confiscated because the size of the animals did not match what was declared on the CITES paperwork (TRAFFIC, 2011). Similarly, 98 Mastigure lizards were taken because they were hatchlings when the documentation indicated they were 18 months old (TRAFFIC, 2011). Finally, 100 Bell's hinged tortoises were seized as officers saw indications of them being wild-caught rather than ranched as was written on the CITES export permit (TRAFFIC, 2011). Then in 2009, agents seized 26 royal pythons from the United States that did not have the correct import paperwork (TRAFFIC, 2011).

According to staff of the CITES enforcement team at Heathrow (personal communication, 26 June, 2011) currently for animals it is the tortoises that are causing the most issues and *Aquilaria* species (agarwood) in terms of plants and timber. There is no real single country that is the source of the illegal trade, although the North African, Southern African and non-EU Mediterranean regions are causing concern in respect to tortoises (personal communication, CITES enforcement officer, 26 June 2011). In 2011, the interviewee led two challenging investigations into the illegal importation of 200 tortoises. The investigation both within the United Kingdom and the country of origin uncovered that the tortoises were not captive-bred, leading to their seizure for being illegally imported (personal communication, CITES enforcement officer, 26 June 2011).

In addition to reptiles, there have also been incidents of smuggling of birds discovered at Heathrow. In 1998, a shipment originating in Singapore going to Mexico containing a variety of birds and 76 cockatoos of various species had paperwork that indicated that the cockatoos were from Indonesia and were obtained from between 1985 and 1993 (TRAFFIC, 2011). On closer inspection by a veterinarian it was determined that the individuals were much younger than this paperwork indicated (TRAFFIC, 2011). This made the documentation invalid and all the cockatoos were seized and then placed into quarantine as they suffered from psittacosis (TRAFFIC, 2011), a zoonotic disease producing pneumonia-like symptoms in humans. Since HARC has limited space to house confiscated animals, the other birds were sent on to Mexico and the cockatoos were found homes in breeding programs (TRAFFIC, 2011). Smuggling of rare birds has resulted in the longest sentence ever given for a wildlife trade offence (TRAFFIC, 2011). In 2000, two people arrived from Bangkok with 23 rare birds, mostly owls and eagles, placed and taped into plastic tubes within their luggage. Six of the birds had died and the others were in poor health. All were temporarily housed at HARC before being found new homes, as they could not be returned to the wild (TRAFFIC, 2011). The man meeting the couple to collect the birds was given six and a half years for smuggling, possessing, and trading these species (TRAFFIC, 2011). More recently in 2009, unclaimed luggage was found to contain nine dead Houbara bustards (a species in demand for use in falconry), supposedly from a conservation centre in Morocco (TRAFFIC, 2011). Reported seizures of other species, such as fish and mammals do take place at Heathrow, but with less frequency.

Whilst most of the above incidents came to the attention of the authorities for fraudulent or missing paperwork, there were also elements of harm within some of these cases. Tortoises in suitcases, crocodiles with hooks in them, birds of prey stuffed into tubes, and deaths are indications of the harm that is occurring in the illegal wildlife trade. These are violations of species justice, where individual animals suffer and die for the sake of human consumption. The above has provided an overview of the harmful nature and extent of the smuggling and accompanying inspections, as well as the tactics employed by traffickers. HARC, a local agency, plays a key role in uncovering this transnational green crime. The next section will contextualise further the local implications of wildlife trafficking.

The local implications

HARC provides a telling example of how the international legal and illegal wildlife trades can affect local areas that would appear to be far removed from such activity. By deconstructing HARC's activities and the history of wildlife seizures it has been a part of, the significance to local communities begins to become apparent. Further exploration of two detailed instances of confiscations at HARC solidifies the dynamics of power, harm, and injustice that occur on the local level. These can be seen in the financial, security, and animal welfare impacts for local communities, as will be discussed below.

Captivity with no end

In late 2009, HARC staff were inspecting a shipment of ten monitor lizards, which are a CITES Appendix II species. This means that they are not currently threatened with extinction, but that unregulated and excessive trade could drive them towards this. As an Appendix II species, export of monitor lizards requires an export permit and, as a CITES shipment, HARC staff routinely inspected the container with the monitor lizards. Each lizard was in a cloth bag with a footprint of no more than an A3 (16.5 x 11.7 inch) sheet of paper (personal communication HARC staff, 20 April 2011). The lizards were only transiting through HARC on their way from Tanzania to Mexico, yet this limited amount of space with no airholes was a welfare violation, thus making their permits invalid, and HARC staff seized all ten lizards. The importer was contacted and given the opportunity to transport them in compliance with proper welfare standards, but he made no attempt to recover them (personal communication HARC staff, 20 April 2011). Eight of the monitor lizards have been rehoused in zoos or wildlife refuges, but two who have not found facilities willing to take them have been at HARC for over a year and half and will remain there until new housing can be found (personal communication HARC staff, 20 April 2011). While conditions at HARC are adequate, the animal holding areas are not designed to be long-term residences. The small concrete rooms have the bare minimum to care for the animals.

Owner's rights over nonhuman animal's rights

In approximately late 2008, a man was stopped at Heathrow Airport for transporting two ring-tailed lemurs without the proper paperwork (personal communication HARC staff, 20 April 2011). Ring-tailed lemurs are a CITES Appendix I species, which means trade is closely regulated and monitored, as the species faces extinction. For trade to take place, the shipment must have both an export permit from the country of origin and an import permit – he had neither of these (personal communication HARC staff, 20 April 2011). The UKBA CITES enforcement team confiscated the lemurs and housed them at HARC. The man was charged for transporting CITES species without the proper clearance and the court case is ongoing. Since the case is ongoing and the lemurs are considered to be the man's property, the lemurs cannot be relocated (personal communication HARC staff, 20 April 2011). They are housed in a small closet-like cage with limited access to the outdoors. The pair has had four babies while housed at HARC also, which must remain confiscated until the outcome of the trial. There is still the possibility that they will be returned to the man after the case has been settled, but until then the lemurs will remain in limbo and continue to breed.

Financial impacts

As indicated above, HARC houses tens of thousands of legally traded animals every year for varying time periods, as well as illegally traded ones. It might be a few hours until the importer or owner picks up their shipment or, as shown in the above sections describing the monitor lizards and ring-tailed lemurs, in some instances of illegal trading it might be indefinitely – a significant ongoing cost. Housing this many animals costs thousands of pounds annually in facilities and infrastructure. There needs to be proper facilities for the range of species, so fish, large and small mammals, reptiles, amphibians, and insects, all of which require particular conditions, need to have spaces where they can be housed. Birds add an additional layer of cost, as they must all be quarantined for a specific period of time to ensure they are disease free. In regard to facilities, this requires a separate isolated housing block with additional precautions, such as masks and ventilation.

While the animals are being housed, they obviously must be cared for. This entails costs for food, water, and bedding. This means staff costs for the animal care workers that feed the animals; clean the rooms, cages, and aquariums; and check the animals' well-being regularly. Searching for smuggled or stowaway animals and conducting welfare inspections, also means paying the salaries of HARC staff and the UKBA CITES enforcement team. Further staff costs arise from veterinarian care that is needed by the animals. Since HARC plays a law enforcement and regulation enforcement role, it is not allowed to make a profit on its activities (personal communication HARC staff, 20 April 2011). Therefore, when a council, charity, or the UKBA need HARC's help in housing an animal, HARC charges rent for this, but it is just enough to cover their costs (personal communication HARC staff, 20 April 2011). So HARC's operations are partly funded by the services that they provide, but in general funding is through the City of London. As is evident, international wildlife trafficking has significant financial implications for the United Kingdom, the City of London, and ultimately for the tax payers who fund their local authority.

Security impacts

As Karesh et al. (2005) have found, international wildlife trade is one of the main vectors for transmitting diseases. This is not only diseases between animals, but also zoonotic diseases, such as mad cow disease and severe acute respiratory syndrome (SARS), which can transfer from animals to humans. London, and Heathrow Airport, as a main transit point in the world for wildlife must then contend with the fact that both the legal and illegal wildlife moving in and out may carry diseases that can infect other wildlife, animal industries, and people. The task of preventing such transmissions and outbreaks then falls to HARC as the agency that is responsible for health and welfare inspections. The local implications are connected to the above-mentioned financial impacts. There are specific facilities and staff required to implement thorough inspections of the health and documentation of the legal wildlife that is being transported through HARC, as well as facilities and staff that can accommodate an animal that needs to be quarantined. Additionally, there have to be sufficient resources to search shipments for illegally smuggled wildlife that may carry disease and unintentional stowaways, which also can pose a threat. Again, if such an animal were found, HARC must have the proper space to keep a diseased animal and/or the proper equipment to euthanise the animal and destroy the carcass.

There is also the more obvious security issue of keeping wild animals from escaping. Wild animals would potentially pose not only a danger to the public and disruptions to an urbanised area; escaped alien species can pose significant threats to the health of ecosystems. Invasive species can outcompete local species, thus endangering the local wildlife and inflicting environmental harm. The United Kingdom is no stranger to this occurrence, as is evident by the grey squirrel, which is an invasive species that has brought about severe population decline of the native red squirrel. So the HARC building must ensure that animals remain secure inside to protect the local, native area animals and people from disease and invasive species, and to also ensure disease is not passed between the varieties of animals that they are caring for at any one time.

Animal welfare

HARC is the regulating agency for animal health and welfare during transportation checks on the well-being of thousands of animals a year in transit through Heathrow. This makes them responsible for preventing nonhuman animal harm and on some level ensuring species justice. When staff discover that an animal has been shipped in improper conditions or is unwell, they act on behalf of the individual animal to reduce its suffering. In addition to addressing harm and justice, this also has implications for the local area. Violations of the welfare standards set out by international agreements and violations of airline guidelines means that HARC can and does confiscate animals when these standards are not met. Again, this ties into financial implications, because of the housing and staff that are required in such instances. Additionally, there is impact on the animals' welfare. Some animals may never recover from being smuggled or shipped in poor conditions. The lack of food and/or water, extreme temperature fluctuations, trauma, and possible injuries that animals can suffer when transported or smuggled in improper containers can require long-term care and recovery. In some cases, smuggled animals will be unable to be returned to the wild because of the injuries that they have suffered while being smuggled, as indicated above with the caiman crocodiles.

As both detailed examples also indicated, confiscation has long-term impacts on animal welfare, because of the uncertainty of finding new homes for seized animals. In the instance of the monitor lizards, they were being shipped in a way that violated welfare standards; they did not have enough room in the burlap sacks and did not have airholes enabling them to breathe properly. This is a clear instance of harm, which resulted in confiscation of the lizards and 2 of the 10 remaining indefinitely at HARC. From an animal welfare stand point, this is better than the conditions that they were forced to endure during transportation, but in the long term this is not ideal. HARC facilities are not designed for permanent occupation. It is a building designed for animals in transit. And there is the local long term implication of housing and care that must continue for an indefinite period of time.

The ringtail lemur case has similar harm and justice dynamics. The lemurs too are indefinitely living out their lives in a concrete cage with limited access to the outdoors, so they also are in not ideal conditions. They also are continuing to breed, which is quite problematic, as it perpetuates the situation onto other individual animals. The lemurs under species justice deserve to be housed in much better conditions, but because the owner still has a claim on them as his property, they will remain in this barely adequate space until the legal issues are resolved. This highlights the power dynamics that can occur in cases of wildlife trafficking. Ownership rights by humans over another species outweigh consideration of the species' right to live free from harm and injustice. This is arguably a case where animal welfare should be the priority over human claims, but the criminal justice system and mainstream criminology have historically only viewed animals as property (Beirne, 2007). This is a clear case, as argued from a green criminological perspective, when such an inhumane and unjust practice should be challenged.

The United Kingdom as an entire country is certainly a transit country for the illegal trade in live animals and, more specifically, Heathrow is a transit point due to the airline routes (personal communication CITES enforcement officer, 26 June 2011). Not to be overlooked is that the United Kingdom can also be classified as a consumer country for illegally traded species (personal communication CITES enforcement officer, 26 June 2011). But a distinction needs to be made between deliberate smuggling and documentation errors, (personal communication CITES enforcement officer, 26 June 2011) and the UKBA CITES enforcement team come across both, after which HARC will house the animals that are seized. Either intentionally illicit or accidental, there are local financial, security, and animal welfare implications that are concentrated in the London area, though the crime or mistake may have taken place thousands of miles away. Yet regulatory agencies add to the harm and injustice of the illegal wildlife trade by keeping animals in facilities meant to be temporary for indefinite lengths of time out of regard for the property rights of people.

Conclusion

International wildlife trade in its legal and illegal forms impacts upon the local area outside of London and on London itself, because of the unusual oversight the City of London local authority has over HARC. The implications are multifaceted. There are significant financial impacts because of the specialised facilities and staff that are required to fulfil the task of checking the health and welfare of animals being transported through Heathrow Airport, as well as conducting inspections to uncover smuggled illegal wildlife and stowaway animals. Trying to uncover smuggling entails searching for hidden and undeclared animals within legal shipments, and checking that documentation has not been forged or has not misdeclared the species, age, country of origin, or captive status of the animal being transported. With the global scope of the illegal wildlife trade, this requires diligence for all shipments, not just those from certain countries or regions. Other impacts are that may happen with legal and illegal wildlife trade can be ways in which diseases that pass among animals or from animals to humans can be transmitted, so security precautions are essential to HARC's mission. This also means keeping contained all of the wildlife that enters the United Kingdom at Heathrow Airport in order to ensure the safety of the public, to protect the native ecosystems, and to prevent environmental harm. The animals themselves are affected the most directly. Forced to at times endure inhumane conditions during transportation or smuggling, animals can suffer injuries or death during transport. Some animals that are seized live at HARC for the foreseeable future because either no permanent facility is willing or able to take them or because the criminal justice process leaves their future in an uncertain state. This exposes the power of humans over their animal property, and the harm and injustice on the individual level that victims of the illegal wildlife trade are forced to suffer. All these implications are reason to improve efforts to stop the illegal wildlife trade, be it for money, safety and health, or animal welfare concerns. And as I hope I have demonstrated here, this isn't just happening on the other side of the world, but at an airport or port nearby.

References

Animal Health Act 2002 (UK). http://www.legislation.gov.uk/ukpga/2002/42/contents.

Beirne, P. (2007) 'Animal rights, animal abuse and green criminology', in P. Beirne and N. South (eds) *Issues in Green Criminology: Confronting Harms against Environments, Humanity and Other Animals*. Devon, UK: Willan Publishing, pp. 55–86.

City of London (2012) *Heathrow Animal Reception Centre*. Available at http://www.cityoflondon.gov.uk/Corporation/LGNL_Services/Environment_and_planning/Animal_welfare/Heathrow_animal_reception_centre.htm. Accessed 13 February 2012.

Convention on the International Trade in Endangered Species of Wild Fauna and Flora (CITES) (1979) *Advice to Carriers*. Available at http://www.cites.org/eng/resources/transport/advice.shtml. Accessed 23 February 2012.

Convention on the International Trade in Endangered Species of Wild Fauna and Flora (no date) *What is CITES?* Available at http://www.cites.org/eng/disc/what.php. Accessed 13 February 2012.

Dangerous Wild Animal Act 1976. Available at http://www.legislation.gov.uk/ukpga/1976/38. Accessed 23 February 2012.

Department of Environment, Food and Rural Affairs (DEFRA) (2007) *Customs CITES Seizures 2007*. Available at http://archive.defra.gov.uk/wildlife-pets/wildlife/trade-crime/documents/custom-seizure2007.pdf. Accessed 23 March 2012.

Department of Environment, Food and Rural Affairs (DEFRA) (2011) Implementation of the Live Animals 'Balai' (Catch-all) Directive in Great Britain: Council Directive 92/65/EEC (as amended). Available at http://animalhealth.defra.gov.uk/about/publications/balai/balai-directive-introduction-general-requirements.pdf. Accessed 22 February 2012.

Karesh, W., Cook, R., Bennett, E. and Newcomb, J. (2005) *Emerging Infectious Diseases*. Available at http://www.cdc.gov/ncidod/EID/vol11no07/05–0194.htm. Accessed 20 May 2011.

TRAFFIC (2011) TRAFFIC Bulletin Seizures and Prosecutions March 1997–April 2011. Cambridge, UK: TRAFFIC International.

Welfare of Animals during Transport Order 1997 (UK)

White, R. (2011) *Transnational Environmental Crime: Toward an Eco-global Criminology*. London: Routledge.

World Wide Fund for Nature (WWF) (2002) *One Million Illegal Wildlife Items Seized at UK Ports and Airports*. WWF Press release February 2002.

Part III

Policing, Regulation, and Enforcement

7
Perspectives on Criminality in Wildlife

Angus Nurse

Introduction

This chapter examines criminality in wildlife crime, a distinct aspect of green criminology (Beirne and South, 2007; Lynch and Stretesky, 2003) within animal abuse (Henry, 2004; Linzey, 2009) and species justice discourse (White, 2008). The legal protection afforded to animals is socially constructed, influenced by social locations, power relations in society, and the need to both promote and protect specific ideological positions on animals by legislators and policymakers. Attitudes towards wild animals both on the part of offenders who harm them and the society that punishes them, or in some cases allows the harm to continue, reveal much about tolerance for different forms of violence within society, sympathy towards the suffering of others, the capacity for empathy (Beetz, 2009), or an inclination towards violence or other forms of antisocial behaviour (Linzey, 2009).

Benton (1998: 149) suggests that 'it is widely recognized that members of other animal species and the rest of nonhuman nature urgently need to be protected from destructive human activities'. However criminal law predominantly treats animals as property, failing to recognise animals' status as victims or to extend thinking about animals beyond traditional human ideals of justice as a punitive or rehabilitative ideal and failing to incorporate shared concepts of reparative and restorative justice between humans and nonhuman animals. Wildlife crime policy generally treats all offenders as rational profit-driven actors, while public policy statements often fail to identify wildlife crime's causes and fail to clarify the intended impact of enforcement policy beyond basic ideas of detection or apprehension. This chapter's consideration of offender behaviour, its causes, and the United Kingdom's policy response explicitly considers distinct aspects of criminal behaviour and what the abuse and exploitation of wildlife reveals about criminal personalities, motivations and behaviour. It examines perspectives on wildlife criminality, identifying specific types of offenders, discussing their criminality in some detail and making recommendations on dealing with wildlife offenders.

Identifying the wildlife offender

Nurse (2003, 2009 and 2011) examined evidence on wildlife crime policy, criminality, and the enforcement of UK wildlife law between 2001 and 2009, with subsequent reviews in 2010 and 2011. Primary and secondary sources were used to obtain both factual data on policy perspectives and qualitative data on the perceptions of criminality and the effectiveness of UK wildlife law enforcement. Semistructured interviews with wildlife crime practitioners, policy makers, and researchers were undertaken to include representatives of the leading UK wildlife crime nongovernmental organisations (NGOs): the Royal Society for the Protection of Birds (RSPB), the Royal Society for the Prevention of Cruelty to Animals (RSPCA) and its Scottish equivalent, the Scottish Society for the Prevention of Cruelty to Animals (SSPCA), the League Against Cruel Sports (LACS) and Scottish Badgers, plus selected police and other statutory enforcement representatives. Interviews were supplemented with documentary analysis of published policy perspectives, media releases and campaign material, transcripts of cases, and submissions to government on wildlife crime issues.

Wildlife crime, loosely defined in this chapter as any act prohibited under UK or European Union (EU) law and involving or targeted at wild birds, mammals, or other animals, predominantly involves the exercise of power by the dominant over the more vulnerable. Groombridge (1996) identifies crime as a predominantly male concern, reflecting the importance of gender and predominance of male offenders in serious and violent crime. The socialisation of young men and the extent to which routes to manhood leave young men confused or anxious about what it means to be a man can influence young males' criminality (Harland, Beattie and McCready, 2005; Kimmell, Hearn and Connell, 2005). Restrictive notions of masculinity dictate that many men are forced into roles as defenders and protectors of their communities

(Harland et al., 2005) and are also encouraged to comply with the image of the 'fearless male' (Goodey, 1997: 401) and achieve the ideal of hegemonic masculinity (Connell, 1995; Harland et al., 2005). Many wildlife crimes involve appropriate male behaviours such as aggression, thrill-seeking, or having an adventurous nature, which reinforces hegemonic masculinity (Harland et al., 2005). Recklessness and assertiveness are conducive to committing wildlife crime in sometimes difficult and dangerous outdoor conditions, with a requirement to negotiate wildlife (for example, dangerous species and adult wildlife protecting its young) and the attentions of law enforcement and NGOs. In addition, the aggressive nature of such crimes as badger baiting (and digging), and hare coursing, and the opportunities for gambling related to these offences (and others, such as cock fighting) appeal to young men seeking to establish their identity and assert their masculinity and power over others. Such crimes by their very nature provide opportunities for men to engage in and observe violence (Flynn, 2002) and to train animals (fighting cocks, dogs) that represent an extension of themselves and reinforce elements of male pride, strength, endurance, and the ability to endure pain.

Wise (2000) argues that the concept of inequality between humans and nonhumans is central not just to the legal status of animals, but also to how individuals treat animals. The perception that certain animals do not feel pain allows offenders to commit their offences without considering the impact of their actions or feeling any guilt over them. Denial of injury is an important factor; it not only allows individuals to deny the harm caused by their actions, but also perpetuates the view of animals as a commodity, rather than as sentient beings suffering as a result of individual criminality, and it allows individuals denial of this criminality.

Attitudes towards regulation are also an important factor in defining wildlife offending. Eliason's (2003) assessment of poachers in Kentucky (which defined *poaching* as the illegal taking of wildlife resources, comparable to UK wildlife crime classifications) concluded that convicted poachers routinely employed neutralisation techniques. These techniques included denial of responsibility, claim of entitlement, denial of the necessity of the law, defence of necessity, and recreation and excitement (Sykes and Matza, 1957), both before and after engaging in illegal activity. Significant numbers of those interviewed by Eliason were aware that they were contravening regulations, but considered that their breaches were minor or technical infringements that should not be subject to law enforcement attention. They often also denied the legitimacy of law enforcement action against them, contending that there were better uses of officers' time – i.e., enforcement action directed towards 'real' criminals.

The involvement of environmental NGOs, without which wildlife offenders might not be apprehended, provides additional motivation for some individuals to commit crime. By considering the different motivations, neutralisations, and behaviours of offenders, it is possible to determine distinct types of wildlife offender, as described below.

Developing offender models

Despite criminal justice policy bias towards profit-driven rationality, wildlife criminality is varied (Nurse, 2011) and is driven by the following general reasons:

1. Profit or commercial gain
2. Thrill or sport
3. Necessity of obtaining food
4. Antipathy towards governmental and law enforcement bodies
5. Tradition and cultural reasons

While these are the primary motivations, ignorance of the law is also sometimes a factor, although it is not strictly a motivating factor, but more a justification or neutralisation technique (Sykes and Matza, 1957). Wildlife offenders fall into four relatively distinct types, defined by their primary motivator, as follows:

A. **Traditional Criminals**, who derive direct and sometimes personal financial benefit from their crimes.
B. **Economic Criminals**, who commit wildlife crimes as a direct result of particular economic pressures – for example, direct employer pressure or profit-driven crime within their chosen profession. This category is distinguished from the previous category because of the specific, mostly legitimate, employment-related nature of their motivation to commit crime.
C. **Masculinities Criminals**, who commit offences involving harm to animals, exercising a stereotypical masculine nature both in terms of the exercise of power over animals and links to sport and gambling. There is some link between these offences and low-level organised crime.

D. **Hobby Criminals**, who commit low-level crimes for which there is no direct benefit or underlying criminal 'need' and for which the criminal justice reaction is often disproportionate. These are distinguished from the previous category by the absence of harm/cruelty as a factor in the offences. The 'hobby' element is the primary motivator.

Within these categories a range of perspectives on power, harm, and justice are evident on the part of offenders, in the public policy response to their crimes, and in law enforcement activity.

Model A – the traditional criminal

Model A criminals obtain personal financial benefit from their actions; wildlife is simply a commodity through which this primary motivator may be achieved and could be substituted for any other activity or commodity providing comparable financial benefit. Model A offenders include: those who take wild bird chicks for breeding and subsequent sale as falconry birds, those who deal in illegally killed wild birds or animals, and wildlife traders in rare or endangered species. This profit-driven crime reflects the absence of more acceptable means of wealth acquisition as both offender motivation and offender engagement in particular wildlife activities. Model A thus represents a rational-choice offender, most likely unaware of the full extent of wildlife legislation, but aware that his or her actions are unlawful in some way.

Rationalisation and determination

Opportunity (Clarke, 1992) and an easy source of direct financial gain are two of the causes of Model A crime. Wildlife crime presents a low-risk, high-return option for the offender, with the potential to make thousands of pounds in a single transaction, and the risks of detection, apprehension, and punishment are slight when compared with other offences. Traditional wildlife offenders also rationalise their crimes by viewing wildlife as an available resource and one over which man routinely exercises power; thus their offences are (to them), at best, minor or technical crimes.

Wildlife resources are not closely monitored by criminal justice agencies. Nor are they the subject of intensive crime prevention or target hardening initiatives employed to protect other valuable commodities. Except in the case of rarer species, the nests of wild birds are not routinely monitored, and only certain birds are required to be registered with the Department for the Environment Food and Rural Affairs

(DEFRA) under current UK legislation. Thus the situational crime measures used as a deterrent in other forms of crime are absent in wildlife crime (Nurse, 2009; Wellsmith, 2010 and 2011). Model A offenders are likely aware that wildlife enforcement is carried out predominantly by NGOs and that wildlife crime penalties are comparatively small.

In addition, the relatively low stigma generally attached to wildlife crimes (i.e., that they are not widely categorised as 'serious' crime) allows offenders to rationalise their offending behaviour as harmless, technical, or victimless offences. With the exception of organised crime's involvement in wildlife crime, where criminal gangs have diversified into wildlife crime using the same routes employed for trading in heroin and cannabis (House of Commons Environmental Audit Committee, 2008; The Scottish Institute for Policing Research, 2007), individual Model A offenders are not classified by society as dangerous or serious criminals. There are, however, animal policy advocates who contend that they should be viewed as such, given the failure of current approaches to wildlife crime to prevent animal exploitation and harm.

The public policy response

The public policy response to Model A offenders treats the offender as a rational actor choosing his course of action and sufficiently aware of his criminal nature that a deterrent approach might be effective. By raising the offender's level of awareness of the likely punishment, NGOs and criminal justice agencies hope to effect behavioural change. Publicity for convictions and the likely level of punishment are, therefore, an important part of the public policy response (Nurse, 2003). In addition, as offenders come to be considered to be persistent offenders, motivated entirely by profit and personal gain, moves towards a more punitive sentencing regime are advocated for the traditional wildlife offender.

Model B – the economic criminal

Economic criminals are primarily motivated by economic and social pressures, but their primary objective is not personal financial benefit. This category includes those who commit wildlife crimes during the course of their employment, as a result of direct and indirect pressure from employers and other employment stakeholders. Examples include: gamekeepers and others involved in what are mostly legitimate countryside sports, game rearing or commercial fisheries, driven to their offending behaviour through employer interaction. This category also includes offences committed by a company or business in the conduct of an otherwise lawful business, often for commercial reasons, thus

Model B is distinguished from Model A by the otherwise lawful operation of the Model B offender.

The Model B offender's motivation comes from external pressures (employer or perception of market pressures) or association with others within his sphere of employment or social circle who have also committed offences (Sutherland, 1973). In game rearing, for example, evidence from investigations suggests that gamekeepers are encouraged to kill otherwise protected birds, animals, and mammals by employers. The objective is to maximise the availability of game for clients and retain the economic viability of the business. A well-stocked estate is essential to ensure successful shooting days and repeat customers, thus internal, peer, and employer pressure encourage little discrimination between legal target predator species such as foxes (which can legally be shot) and birds of prey (protected at all times). Gamekeepers may be otherwise law-abiding individuals and will frequently cooperate with the police over other crimes such as poaching. Timber treatment staff and building and roofing contractors also feature amongst these offenders, because wildlife survey costs are not justified by the relatively low risk of apprehension, prosecution, and conviction. Thus staff will carry out building works that negatively impact on wildlife and which contravene wildlife legislation. The offender is most likely aware that his actions amount to offences under wildlife legislation but, because of multiple pressures, he continues to commit offences.

These offenders can be likened to white-collar criminals where 'successful business or professional people are apparently caught out in serious offences, quite often for behaviour which they did not expect to be treated as criminal, and for which it is quite difficult to secure a conviction' (Nelken, 1994: 355). Model B offenders are frequently responsible people, for example countryside professionals employed in lawful pest control, who in theory at least stand to lose their jobs and homes if convicted (see below).

Rationalisation and determination

Economic wildlife crime is directly related to outside pressures and lack of controls on offenders' activities. The rationalisations used by offenders differ from those of the traditional criminal and are based not only on economic reasons, but often on arguments contesting limitations on allowable wildlife control and denying the wisdom of the legislation under which offenders must operate.

In the case of companies, Situ and Emmons commented that: 'performance pressure, the estimated certainty and severity of punishment, and the crime facilitative culture at the level of the individual firm contribute to the probability of criminal participation' (2000: 60). The pressure to kill protected wildlife can either be direct or indirect – for example, an employer directly informing the gamekeeper that birds of prey and other predators are to be controlled, or simply turning a blind eye to the activities of a gamekeeper who is regularly producing high levels of game for the estate. Evidence from case files also identifies that some new gamekeepers learn illegal techniques of predator control from other more senior staff (Nurse, 2011). Recognition of the role of the employer in encouraging offences is reflected in attempts to reduce landowner pressure on staff to commit crime and create landowner liability, now partially implemented in the Scottish offence of 'vicarious liability' for wildlife crime.

Economic offenders rationalise their unlawful activities as being the responsibility of others, including the employer who pressures them. Offenders thus deny culpability for their actions, and argue that committing offences is a necessity of earning a living and providing for their family, that their crimes are victimless and of a technical or minor nature, and that the resources of the criminal justice agencies should be targeted towards 'real' and serious criminals.

In part these rationalisations are a defence mechanism against the perception (and campaigning) by NGOs that wildlife crime is serious crime, justifying mainstream criminal justice agency attention, and demonstrate Sykes and Matza's (1957) neutralisations at work. That the responsibility for wildlife law enforcement is largely the responsibility of NGOs is significant. Offenders are aware that the likelihood of getting caught and the likely fines if convicted potentially work in their favour, and commercial interests frequently mean that the potential punishment is obviated by the significant returns that can be achieved by ignoring rather than complying with wildlife laws. Commercial expediency thus provides a rationalisation for the offences, since to fully comply with the legislation costs money, delays projects, and puts the company's profits at risk, while leaving the company vulnerable to competition from a company with a more 'flexible' attitude to wildlife legislation.

The public policy response

The public policy response to economic criminals is variable. Publicly, UK game-rearing estates state that any gamekeeper convicted of a wildlife offence would be dismissed. NGOs argue that this is not the case, and that an offender can continue to commit offences without fear of any further sanctions being applied after conviction. Offences are detected and prosecuted primarily as a result of the efforts of the

NGOs, according to the law enforcement detection and apprehension model, rather than as a result of any concerted effort by any statutory criminal justice agencies.

Model C – the masculinities criminal

Model C offenders are primarily motivated by power and notions of masculinity; such offences are seldom committed by lone individuals. In some crimes, the main motivation is exercising power allied to sport or entertainment; a link might also be made with organised crime and gambling. Such crimes, classed as crimes of masculinities, also include elements of cruelty or animal abuse of the kind that attracts law enforcement agency attention in the United States (Clawson, 2009). Examples include badger digging and badger baiting, and cockfighting, as well as some crimes that involve the 'sporting' killing or taking of wildlife. (This is to be distinguished from legitimate predator control activity or the killing of badgers to prevent the spread of bovine tuberculosis.) Evidence from the RSPCA, SSPCA, and LACS suggests that in these crimes, the offender is likely to derive some pleasure from his offence and this is a primary motivator, and that there is a link between some of these crimes and other crimes of masculinities.

Rationalisation and determination

Badger baiting, cockfighting, and hare coursing are considered by some to be sports, although the inevitability of animal injury is significant. Anti–field sports NGOs conclude that such sports attract individuals specifically attracted by harming animals, by the excitement and enthusiasm of causing such harm, and by engaging in the illegal activity, but note that these sports have a desensitising effect on participants (Morrisey, 2004: 13–17).

American research on wildlife-oriented crimes of the masculine, including cockfighting and cockfighting gangs, identifies cockfighting as having 'a mythos centred on the purported behaviour and character of the gamecock itself. Cocks are seen as emblems of bravery and resistance in the face of insurmountable odds' (Hawley, 1993: 2). The fighting involved is 'an affirmation of masculine identity in an increasingly complex and diverse era' (1993: 1) and the fighting spirit of the birds has great symbolic significance to participants, as does the ability of fighting and hunting dogs to take punishment in UK wildlife crime.

Masculine stereotypes, reinforced and developed through offending behaviour (Goodey, 1997), are important factors in Model C offending. Wildlife offenders in the United Kingdom are almost exclusively male, and the more violent forms of wildlife offender exhibit distinctly masculine characteristics. Evidence suggests that younger wildlife masculinities offenders could turn to more serious forms of crime or expand their violent activities beyond animals and towards humans (Ascione, 1993; Flynn, 2002; Clawson, 2009). Hare coursing, cockfighting, and badger digging all involve gambling, w ith wagers being placed on individual animals, the outcome of a fight, and other factors (including the power or strength of an animal). For some, the associated gambling is as important as the exercise of power; significant sums are waged on fights, attracting the attention of organised crime.

Significantly, some crimes, such as badger digging, are group activities where group relationships replicate informal criminal behaviour (RSPCA, 2006, 2007). A 'secret society' or 'old boy network' exists for wildlife crimes, where individuals can call upon others for collaboration, help, or services when they need them, and are be able to verify their 'bona fides' to those they did not know, and the community actually encourages crime (Maguire, 2000:131). The male-bonding element identified by Hawley (1993) is significant, banding together men from the margins of society and for whom issues of belonging, male pride, and achievement are important. Younger cockfighters 'are taken under the wing of an older male relative or father, and taught all aspects of chicken care and lore pertaining to the sport' (Hawley, 1993: 5); women are generally excluded. Forsyth and Evans (1998) reached similar findings in researching dog fighting in the United States, concluding that an appeal to higher loyalties and an attachment to smaller groups took precedence over attachment to society.

Model C offenders often rationalise based on historical precedent or tradition. Resistance to the UK's Hunting Act 2004, which banned hunting with dogs, employed arguments emphasising the traditional nature of hunting and dismissed antihunting legislation as Whitehall interference in the countryside. Hunting supporters also deny that hunted animals feel pain, and they stress hunting as necessary and effective predator control. Even after the introduction of the Hunting Act of 2004, its proponents continue to challenge its legitimacy. *Jackson v. Attorney General* [2005] UKHL 56 represented an unsuccessful attempt to challenge the Hunting Act's validity on constitutional grounds. Common to Model C offenders are rationalisations based on denial of legitimacy, and claims of unwarranted intervention by legislators and a lack of understanding on the part of those that seek to ban the activity. The Act was also unsuccessfully challenged on the grounds that it was incompatible with the European Convention on

Human Rights in *R (Countryside Alliance and Others) v. Attorney-General and others; R (Derwin and Others) v. Attorney General*, [2007] UKHL 52.

The public policy response

The public policy response to masculinities crimes reflects acceptance of offenders' violent tendencies and is similar to responses to organised crime. Techniques employed by enforcers include infiltration of gangs, surveillance activities, and undercover operations. Masculinities offences are considered to be more dangerous than other criminal wildlife behaviours and are treated accordingly.

Model D – the 'hobby' criminal

Offenders who are involved in technical offences for which they often deny their criminal characterisation constitute the final category. These offences attract a disproportionately high level of attention from criminal justice agencies and NGOs, given their relatively low 'threat' level. Model D offenders include egg collectors, who gain little direct benefit from their offence and for whom the criminality involved is denied. It also includes large-scale taxidermy collectors, who do not operate mainly as traders or dealers.

Model D offenders are driven by the collection or *acquisition* of items. Their offences are not generally committed for business or occupational purposes and can be more readily likened to a hobby or obsession. Egg collectors, for example, are rarely countryside employees, but instead are those employed (or unemployed) elsewhere, who specifically travel to the countryside to commit offences. This element of *mens rea* accounts for the seriousness with which these offences are considered by NGOs, criminal justice agencies, and the public.

Hobby criminals' obsessive pursuit of their hobby can cost thousands of pounds annually; collectors have travelled extensively over Europe in pursuit of eggs and individuals involved in (illegal) taxidermy have been found in possession of species taken globally.

Rationalisation and determination

Hobby wildlife crimes defy comprehensive explanation as generally no financial gain is derived from the activity. However, analysis (RSPB, 1999 and Wainwright, 2006) identifies obsession as a behavioural explanation:

> [Egg collecting] is purely an obsessive and selfish activity resulting in nothing more than displaying the egg in a purpose built cabinet to

> gaze at until the start of the next breeding season, when additions to the collection can be made. (RSPB, 1999: 20)

Egg collecting has been likened to a form of kleptomania or obsessive-compulsive disorder; offenders are driven to commit their crimes and are addicted to the adventure involved in doing so. Similarities are evident with other obsessional offenders, those who collect banned or expensive items like rare books, pornography, and stolen paintings (Burke, 2001; Taylor and Quayle, 2003). The desire not just to obtain items but also to catalogue and categorise them is a significant behavioural factor. Stolen art works and books, many of which are recognisable, cannot be traded on the open market, but are acquired for private collectors to appreciate, with criminal gangs turning to trafficking for private collectors and with thefts of works, for example those of Copernicus and Ptolemy, being commissioned by private collectors (Burke, 2001). The drive to obtain items for personal use that cannot be publicly exhibited is a primary factor of the obsessive collector. Taylor and Quayle (2003: 48) explain that 'the emotional intensity that is part of collecting behaviour' is a significant factor, with the collector interacting within specific communities of kindred obsessives, each driven to have a bigger, better, and more comprehensive collection than others. The competitive drive and obsessive need to acquire items can turn a hobby interest into a passionate, compulsive desire to collect (Belk, 1995; Taylor and Quayle, 2003), as egg collector Derek Lee explained:

> There are quite a few who are obsessed with it. Every single spring and summer they can't wait to get out. If you put a child in a chocolate factory their eyes light up with excitement. It's like that. When spring and summer come, the eggers are on edge. They're like big kids. (Barkham, 2006)

The obsessive-compulsive nature of hobby wildlife offending includes collectors keeping meticulous notes of their activities, which are used by investigators as evidence of criminality (Barkham, 2006; Wood, 2008). Egg collecting is learned from others within the community, with established collectors passing on their knowledge and techniques. Spouses and others within the community may, however, disapprove of the activity and may contact enforcement bodies to provide evidence of offending (Nurse, 2009, 2011).

Hobby wildlife offenders use techniques of avoidance, denial of criminality, displacement of blame, and challenges to the legitimacy of enforcers to explain away their actions. Much like those who are caught speeding by traffic enforcement cameras and challenge the cameras' legitimacy, the fines imposed, or argue that cameras are simply a revenue-raising device (Fylan et al., 2006), hobby wildlife offenders dispute that their activities fall within the remit of the criminal law. Denial of criminality and avoidance of responsibility is an integral part of the offender's rationalisation. Egg collecting was once a schoolboy hobby in Britain, given scientific legitimacy as *oology*. It is only with the introduction of the Countryside and Rights of Way Act 2000 (UK) that offences have carried a limited option for prison sentences, although the taking of wild birds' eggs has been unlawful since the 1950s under the Protection of Birds Act 1954 (UK) and subsequent legislation. The fact that wildlife legislation falls outside mainstream criminal justice strengthens offenders' classification of their activities as minor or victimless crime (Wood, 2008) and their denial of NGO enforcement action legitimacy (Nurse, 2011).

The public policy response

Hobby wildlife offences attract a punitive response that is arguably excessive for the nature of the offence. In the United Kingdom, a number of joint police/NGO operations look into egg collecting, and high-profile convictions for egg collecting that attracted large fines have occurred. Situational crime prevention techniques have been employed; the nests of some rare birds, for example osprey and golden eagle, are routinely watched by volunteer wardens during breeding seasons. Osprey nests in Scotland and red kite nests in Wales have also been watched by the army in the past as part of training exercises and to gain publicity for wildlife crimes. The ease with which egg collectors can be demonised, and the seemingly macabre obsession with which collecting dead specimens can be labelled as 'deviant' makes hobby collecting newsworthy. Thus the relative ease of prosecuting such cases, given that prosecutors often simply need to prove possession: the burden of showing lawful possession rests on suspects and ensures high visibility; cases are routinely prosecuted and attract publicity.

Power and harm: dealing with offenders

An examination of the primary motivations and offending behaviour in wildlife crime shows that, rather than there being one 'rational' wildlife offender committing crimes for profit, several distinct offender types exist.

While the nature of the offences may be different, there is inevitably some overlap in the behaviours of offenders, although the weight attached to various determining factors varies. Egg collectors, badger diggers, and gamekeepers are all, for example, keeping a traditional activity alive, but in different ways and for different reasons. The egg collector is pursuing his 'traditional' hobby, whereas the gamekeeper is perpetuating a learned traditional behaviour in the form of predator control techniques handed down from gamekeeper to gamekeeper, irrespective of changes in the law. The masculinities criminal may derive some financial gain from gambling but gain is not a *primary* motivating factor, whereas money is for the traditional criminal. What all offender types share in common is the likely knowledge that their activities may be illegal (although there may be denial as to whether this should be the case) and that the likelihood of detection, apprehension, and prosecution remains low.

Current policy treats all wildlife offenders as traditional (i.e., rational and financially motivated) criminals. In effect the public policy response for the traditional criminal is employed for all offenders, despite the different motivations and rationalisations shown by other groups. However, the different primary motivating factors indicate that different elements drive offenders, and so there is little point in treating all offenders as if they were the same. One conclusion of this chapter is that a blanket approach to dealing with wildlife crime and offenders is unlikely to be successful and represents a flawed justice model that fails to address illegal wildlife harm and exploitation. The enforcement regime therefore needs to be adapted to provide for action appropriate to the circumstances of the offender and the specific nature of the offence. For traditional criminals, financial penalties may work as a means of negating any benefit they derive from their activity, but the same approach is unlikely to work with economic criminals. An argument can also be made that increased sentencing and use of prison has been unsuccessful in mainstream criminal justice (Wilson, 1985), and so the evidence that it will be effective in reducing or preventing wildlife crime is lacking. For traditional criminals, greater efforts should be made to attempt situational crime prevention (Wellsmith, 2011), making the physical cost of committing the crime prohibitive as well as the actual cost and removing the perception that wildlife crime may be seen as a soft option.

For economic criminals, their employment provides the source of their offending behaviour, and so any policy approach must include pressure on and penalties for the employer as well as action that dictates that the risk of losing employment as a direct consequence of committing a wildlife crime is a real possibility. The current legislative regime does not provide for culpability of landowners/employers for the actions of their staff, nor do countryside and game industry employees suffer the stigma of conviction. As a practical means of dealing with these offenders, this position should be altered so that conviction of a wildlife crime carries with it the threat of lost employment in the countryside and in the game rearing or field sports industries, as well as significant penalties for the employer.

For the masculinities offender, the effectiveness of prison or high fines is also questionable. Much like inner-city gang members, masculinities offenders may come to see prison as simply an occupational hazard, as well as reinforcing their male identity and confirmation of society's lack of understanding of their needs and culture (Nurse, 2009, 2011). For masculinities offenders, situational crime prevention should be attempted and a real effort at rehabilitation made alongside the traditional law enforcement approach of detection and prosecution. Consideration may also need to be given to the circumstances in which groups of young men turn to crime with a violent element, exercising power over the vulnerable, and to whether the type of social work intervention combined with law enforcement activity that now takes place in parts of the United States with animal abusers (Brantley, 2009; Clawson, 2009) could be applied in the United Kingdom.

Hobby offenders present a distinct policy and enforcement challenge, as the drive to collect and the obsessive behaviour of such offenders cannot easily be overcome; fines and prison sentences could even strengthen the desire to offend by creating the drive to replace lost items, such as a confiscated egg collection. While prevention and detection of crimes should continue to be employed for these offenders, treatment to address the issues of collecting as well as education in the effects of wildlife crimes should be considered.

Wildlife criminality represents varied aspects of the exercise of power over the vulnerable (Flynn, 2009). Animal victims rely on NGOs and a justice system that punishes exploitation, recognises harm, and considers the links between wildlife crime and other crimes. Yet, contrary to the assumptions inherent in current policy, offenders do not all share the same motivations, behaviours, or operate within similar communities or control mechanisms. The UK wildlife crime enforcement regime therefore needs to be adapted to provide for appropriate action that fits the circumstances of the offender and allows the specific nature of the offence to be taken into account.

References

Ascione, F. R. (1993) 'Children who are cruel to animals: a review of research and implications for developmental psychopathology', *Anthrozoos*, 4: 226–227.

Barkham, P. (2006) 'The egg snatchers', *The Guardian*, 11 December.

Beirne, P. and South, N. (eds) (2007) *Issues in Green Criminology: Confronting Harms against Environments, Humanity and Other Animals*. Devon: Willan.

Beetz, A. (2009) 'Empathy as an indicator of emotional development', in A. Linzey (ed.) *The Link between Animal Abuse and Human Violence*. Eastbourne: Sussex Academic Press.

Belk, R. W. (1995) *Collecting in a Consumer Society*. London: Routledge.

Benton, T. (1998) 'Rights and justice on a shared planet: more rights or new relations?' *Theoretical Criminology*, 2(2): 149–175.

Brantley, A. (2009) 'An FBI perspective on animal cruelty', in A. Linzey (ed.) *The Link between Animal Abuse and Human Violence*. Eastbourne: Sussex Academic Press.

Burke, J. (2001) 'Britain leads illicit trade in rare books', *The Observer*, Sunday 10 June.

Clarke, R. V. (1992) *Situational Crime Prevention*. New York: Harrow and Heston.

Clawson, E. (2009) 'Canaries in the mine: the priority of human welfare in animal abuse prosecution', in A. Linzey (ed.) *The Link between Animal Abuse and Human Violence*. Eastbourne: Sussex Academic Press.

Connell, R. W. (1995) *Masculinities*. Oxford: Blackwell Publishers.

Eliason, S. L. (2003) 'Illegal hunting and angling: the neutralization of wildlife law violations', *Society and Animals*, 11(3).

Flynn, C. P. (2002) 'Hunting and illegal violence against humans and other animals: exploring the relationship', *Society and Animals*, 10(2).

Flynn, C. P. (2009) 'Women-battering, pet abuse, and human-animal relationships', in A. Linzey (ed.) *The Link between Animal Abuse and Human Violence*. Eastbourne: Sussex Academic Press.

Forsyth, C. J. and Evans, R. D. (1998) 'Dogmen: the rationalisation of deviance', *Society and Animals*, 6(3).

Fylan F., Hempel, S., Grunfeld, E. A., Connor, M. and Lawton, R. (2006) 'Evidence-based behavioural change for speeding drivers', *Behavioural Research in Road Safety*.

Goodey, J. (1997) 'Masculinities, fear of crime and fearlessness', *The British Journal of Sociology*, 37(3): 401–418.

Groombridge, N. (1996) 'Masculinities and crimes against the environment', *Theoretical Criminology*, 2(2).

Harland, K., Beattie, K. and McCready, S. (2005) *Young Men and the Squeeze of Masculinity: The Inaugural Paper for the Centre for Young Men's Studies*. Ulster: Centre for Young Men's Studies.

Hawley, F. (1993) 'The moral and conceptual universe of cockfighters: symbolism and rationalization', *Society and Animals*, 1(2).

Henry, B. C. (2004) 'The relationship between animal cruelty, delinquency, and attitudes toward the treatment of animals', *Society and Animals*, 12(3).

Kimmell, M., Hearn, J. and Connell, R. W. (2005) *Handbook of Studies on Men and Masculinities*. London: Sage.

Linzey (ed.) (2009) *The Link between Animal Abuse and Human Violence*. Eastbourne: Sussex Academic Press.

Lynch, M. J. and Stretesky, P. B. M. (2003) 'The meaning of green: contrasting criminological perspectives', *Theoretical Criminology*, 7(2).

Maguire, M. (2000) 'Researching "street criminals": a neglected art', in R. D. King and E. Wincup (eds) *Doing Research on Crime and Justice*. Oxford: Oxford University Press.

Morrisey, C. (2004) *Time to Deliver the Ban*. London/Horsham: RSPCA/IFAW /LACS.

Nelken, D. (1994) 'White collar crime', in M. Maguire, R. Morgan and R. Reiner (eds) *The Oxford Handbook of Criminology*. Oxford: Oxford University Press.

Nurse, A. (2003) *The Nature of Wildlife and Conservation Crime in the UK and its Public Response*. Working Paper No. 9, Birmingham: University of Central England.

Nurse, A. (2009) 'Dealing with animal abusers', in Linzey (ed.) *The Link between Animal Abuse and Human Violence*. Eastbourne: Sussex Academic Press.

Nurse, A. (2011) 'Policing wildlife: perspectives on criminality in wildlife crime', *Papers from the British Criminology Conference*, 11: 38–53.

Royal Society for the Prevention of Cruelty to Animals (2006) *Annual Review 2005: Royal Society for the Prevention of Cruelty of Animals*. Horsham: RSPCA.

Royal Society for the Prevention of Cruelty to Animals (2007) *Annual Review 2006: Royal Society for the Prevention of Cruelty of Animals*. Horsham: RSPCA.

Royal Society for the Protection of Birds (1999) *Birdcrime '98: Offences against Wild Bird Legislation 1998*. Sandy: RSPB.

Scottish Institute for Policing Research (2007) *Forensic Investigation of Wildlife Crime*. Dundee: University of Dundee.

Situ, Y. and Emmons, D. (2000) *Environmental Crime: The Criminal Justice System's Role in Protecting the Environment*. Thousand Oaks: Sage.

Sutherland, E. H. (1973) 'Development of the theory', in K. Schuessler (ed.) *On Analysing Crime*. Chicago: University of Chicago Press (Original work published 1942), pp. 13–29.

Sykes, G. M. and Matza, D. (1957) 'Techniques of neutralization: a theory of delinquency', *American Sociological Review*, 22: 664–673.

Taylor, M. and Quayle, E. (2003) *Child Pornography: An Internet Crime*. London: Brunner-Routledge.

Wainwright, M. (2006) 'The day Britain's most notorious egg collector climbed his last tree', *The Guardian*, 27 May. Available at: http://www.guardian.co.uk /uk/2006/may/27/topstories3.mainsection. Accessed 28 August 2012.

Wellsmith, M. (2010) 'The applicability of crime prevention to problems of environmental harm: a consideration of illicit trade in endangered species', in R. White (ed.) *Global Environmental Harm; Criminological Perspectives*. Cullompton: Willan Publishing, pp. 132–49.

Wellsmith, M. (2011) 'Wildlife crime: the problems of enforcement', *European Journal on Criminal Policy and Research*, 17(2): 125–148.

White, R. (2008) *Crimes against Nature: Environmental Criminology and Ecological Justice*. Devon: Willan.

Wilson, J. Q. (1985) *Thinking about Crime* (2nd edition). New York: Vintage Books.

Wise, S. M. (2000) *Rattling the Cage: Towards Legal Rights for Animals*. London: Profile.

Wood, A. (2008) '"Evil" thief is jailed over haul of 7,000 bird eggs', *Yorkshire Post*. Accessed on 4 June 2008.

8
Environmental Regulation in Chemical Corporations: Preliminary Results of a Case Study

M. H. A. Kluin

Introduction

Incidents like Bhopal, Texas City Refinery, and Deepwater Horizon show us that the consequences of work done by inspectors of chemical corporations is of extreme importance and affects daily life. It is in the public's interest to get more insight into how these street-level bureaucrats arrive at their conclusions. How do they make these decisions? What factors are relevant to the practical definition of compliance? How is compliance defined by these inspectors in their routine work? The focus of this study is on the implementation of regulations through enforcement by field-level inspectors from three different inspectorates: the Environmental Protection Agency, the Occupational Safety and Health Inspection Agency, and the Fire Department. Hutter calls them gatekeepers of the regulatory process, since they are on the front line and determine how regulation is translated into action (1988). These individuals are responsible for the execution of policy and, according to Pautz (2010), for its success and failure.

Seveso inspections in the Netherlands are the objects of this research study. These inspections have been performed for 10 years under the auspices of the Seveso Directive on Major Hazards within the European Union (EU). The inspections are carried out annually by inspectors from the Environmental Protection Agency, the Occupational Safety and Health Inspection Agency, and the Fire Department in the Netherlands. Despite these inspections, there are still many mistakes, accidents, and even deliberate incidents in the chemical industry. Accidents can be organisationally or systematically created and could be viewed as crimes. This does not mean that each and every industrial accident is the result of organisational crime (Tombs and Pearce, 1998: 152). The attention of the current research is devoted to violations of environmental regulations by chemical corporations. One does not immediately associate this subject with criminals. Violations such as these do not fit within the normal range of violations according to the definition of crime in normal literature, or in press and crime statistics. Recent publications on the regulation of environmental violations show that these violations have typical characteristics (Kagan, Gunningham and Thornton, 2011; Simpson and Rorie, 2011).

There is little specific study of the role of field-level inspectors in the implementation of environmental regulations. As Pautz (2010) did, this research begins to respond to this gap by focusing on a small subsection of field-level inspectors in environmental policy, environmental inspectors in the Netherlands. It is important to note that this study is broader than the activities of an isolated group of enforcement officials; it studies the complex social interactive process in which environmental health legislation is implemented. It is a detailed case study of how the Netherlands enforces specific European legislation with a focus on its implementation by field-level inspectors and a focus on compliance by regulatees.

The main purpose of this explorative part of the whole research study is to obtain an understanding of environmental inspectors and their work, to collect more specific data about their working methods, and to establish whether participant observation served as a suitable method of investigation. Understanding these interactions will provide insights into how regulators approach their interactions with regulatees and how they ensure environmental compliance. McCaffrey, Smith and Martinez-Moyano (2007: 308) note that 'understanding how relationships between regulatory organizations and firms affect compliance... is essential to understanding regulator implementation'.

The structure of this chapter is as follows. First, I discuss the theoretical background, the literature of organisational crime, regulatory enforcement, and the Seveso II Directive where relevant. Second, I describe the inspectorate and the practice of environmental inspectors. Before presenting the results of participant observation at four Seveso regulated corporations, I continue by explaining the research method and research setting of this case study. Then I describe the findings of the document analysis and participant observation of four Seveso corporations. In the concluding section of this chapter, the main findings are summarised.

Theoretical background

Relevant literature on organisational crime

Organisational crime is one of the most difficult areas to study within criminology. This is due to its invisibility and complexity, which make it even harder to investigate than common crime. Only a small fraction of violations are discovered and officially recorded. When one starts to investigate crime, most of the time the data are derived from official statistics, crime surveys, and court records. These data do not include organisational crime, only conventional crime statistics, since not many organisational crimes are dealt with by the police. There is a long tradition in research on organisational crime that has its focus on industry case studies, for example: electricity (Geis, 1967); pharmaceutical industries and coal mining industry (Braithwaite, 1984, 1985); Bhopal (Pearce and Tombs, 1998, 2012); criminogenic industries such as the oil, pharmaceutical, and automobile industry according to Clinard and Yeager (1980); and the textile finishing and waste processing industries (Huisman, 2001).

However, the present study is also a study of particular forms of crime. There have been fewer attempts to focus upon particular forms of crime in any specific industry, although in the view of Pearce and Tombs (1998) such work is especially useful of the understanding of the contours of organisational crime. A rare and therefore important example of this kind of research is their own research on corporate crime in the chemical industry. Another important inspiration for my research is Braithwaite's study on the coal mining industry (1985). He studied the organisational differences between coal mining corporations that often violated occupational safety regulations and those that showed great willingness to comply with the regulations. He tried to determine how better compliance and therefore improved safety can be achieved. Braithwaite did his study to gain knowledge of the background of the causes of accidents; he analysed the safety compliance systems of the five corporate leaders at that time in coal mine safety. Hale states similar questions, for example, 'If the regulated [business] cannot be required by law to do no harm, what should they be required to do and what is the role of the regulators? Should they be seeking to punish or persuade?' (Hale, Hopkins and Kirwan, 2002: 3). Braithwaite concludes that disasters in mines can be reduced if there is more willingness to comply with the law. An outcome of Braithwaite's study was that these disasters were most of the time found to be the result of organisational crime. His research shows that a combination of punitive and educative measures taken against offenders can have substantial effects in reducing injuries to miners. Braithwaite found that in the safest coal mine companies, safety was an aspect that was carried out by all the departments of the mine. He concluded that the overall factor in the five safest companies was the approach of the companies to guarantee that all departments comply with safety regulations.

Relevant literature on regulatory enforcement

This research takes a closer look at environmental inspectors, who are examples of street-level bureaucrats. Lipsky (1980) introduced the term 'street-level bureaucrat'; it refers to a public agency employee who actually performs the actions that implement laws. Hutter, whose research was an inspiration as well, says that environmental inspectors are in a position to determine to some extent what constitutes an violation or problem, and moreover to select which cases may be deemed suitable for legal action (1988: 4). 'Regulatory law is often vague, involving broad legal standard and the exercise of discretion by officials' (Hutter, 1997: 3). Field-level inspectors have a great deal of discretion. They need to deal with a wide range of rules and interpret them in real-life situations, and in addition their work is performed in the field in order to make inspections at corporations onsite.

Up until now, much of the study of regulatory actors is focused on the nature of enforcement.

The literature available on regulatory enforcement mentions two types of enforcement styles. This refers to the behaviours of inspectors when interacting with regulated entities (May and Winter, 2011). On one side of the spectrum, an inspector may decide to do 'punitive', 'rule oriented' or 'strict' enforcement (Bardach and Kagan, 1982; Shover et al., 1984) which emphasises a more punitive means of achieving compliance. The other end of the continuum has been labelled a 'compliance' oriented (Hawkins, 1984) or 'accommodative' (Hutter, 1988, 1989) approach, which can be characterised by using more cooperative or conciliatory means. Regulators were thought to operate under one of these approaches; however research demonstrates that they apply these enforcement styles in varying combinations depending on circumstances, settings, and policy areas (Hutter, 1988, 1989; Mascini and van Wijk, 2009a; May and Winter, 2000, 2011). These insights are valuable; however, the focus of attention is on regulatory actors in isolation and it is unable to describe how regulators and the regulatees interact. Interaction of environmental inspectors and regulatees fits within broader discussions of regulation.

According to Hutter (1997) and May and Wood (2003), insight is missing about the regulatory arena and especially about the important role of inspectors at the front lines of regulatory activity. Our present study contributes to the understanding of how those street-level bureaucrats of regulatory enforcement help to bring about compliance with regulations.

In line of the research done by Hutter (1988, 1997) and May & Wood (2003) the focus in this chapter is on interactions between field-level inspectors as street-level bureaucrats for environmental regulatory policy and regulatees.

Before taking a closer look at the inspectorate and the field-level inspectors of the Environmental Protection Agency, this chapter continues with relevant literature on enforcement of the European Union Directive, Seveso II. It is the legal framework in which the inspections take place.

Enforcement of the Seveso II Act in the Netherlands (Council Directive 96/82/EC)

In 1982, the European Union issued chemical industrial safety regulations as a reaction to chemical disasters in England (Flixborough, 1974) and Italy (Seveso, 1976). In Flixborough, an explosion and a fire led to 28 fatalities and the destruction of a plant, and in the case of Seveso, a vapour cloud containing dioxins escaped from a chemical plant. Member states of the European Union negotiated for three years before the Seveso Directive was adopted in 1982 (Versluis, 2003). This directive imposed stringent regulations on the chemical industry. The main purpose of the Seveso regulations is to prevent and mitigate the effects of major accidents involving dangerous substances (Article 1, Seveso II Directive). It monitors consequences of major accidents for the environment and on the other hand it also regulates the protection of employees of a corporation and also people outside the establishment. In the beginning of Seveso, the focus was on the technical part of safety at corporations. Later on it became clear that the causes of many industrial accidents were organisational aspects of safety, such as procedures and communication, especially since the Seveso Directive did not prevent all accidents with dangerous substances, like the one in Bhopal (1984) and in Basel (1986). These accidents outside the European Union led to two amendments of the directive, in 1987[1] and 1988.[2] The first amendment was especially meant to prevent differing interpretations and clarified some entries and threshold levels. After the Basel accident, which caused major environmental pollution of the Rhine, the second amendment focused on the inclusion of isolated storage of dangerous substances.

Since the amendments were considered insufficient to guarantee the aim of preventing major accidents, a complete revision of the directive was recommended by the European Commission and the European Parliament; after long negotiations, the second Seveso Directive was accepted in 1996. It was updated in 1999, amended again in 2005, and is now referred as the Seveso II Directive. A change of focus to more general management systems was one of the main reasons for a completely new directive (Seveso directive II), and this change was too essential to regulate in an amendment, according to Versluis (2003). Seveso II changed other elements as well, compared to the first Seveso Directive:

- the scope was broadened and simplified;
- Seveso II focuses on entire corporations (Article 3, Seveso II Directive) instead of individual installations;
- land-use planning was added;
- there was an extension of the article on inspections, and member states are obligated to enforce the directive (Article 18, Seveso II Directive).

It required operators of major hazard facilities to show the regulator that they have identified, assessed, and controlled the hazards that are present in their facilities.

The setting for this research is the regulation of chemical corporations in the Netherlands; it is called BRZO'99 (Besluit Risico Zware Ongevallen, 1999), the Dutch implementation of the Seveso II Directive. It integrates legislation in the fields of occupational safety, external safety, and disaster management within a legal framework and is an example of a coordination of laws. It represents the policy areas of environment, occupational safety and health, and emergency planning.

Corporations fall under the scope of Seveso II when the quantities of hazardous substances are above a certain permitted level. There are two thresholds, lower tier and upper tier. For each threshold, a limiting amount of dangerous substances is defined. Its objective is to prevent and control major accidents involving dangerous substances. Each corporation has its own responsibility for organizing and securing the safety of its own activities. It is required to have a major accident-prevention policy and a safety management system. If the dangerous substance exceeds an upper-tier threshold, the corporation needs to provide a Safety Report to the authorities as well.

The actual enforcement of the Seveso II Directive is a joint activity of the Netherlands and a Seveso II inspection team made up of three inspectors from different agencies: the Environmental Protection Agency, the Occupational Safety and Health Inspection Agency, and the Fire Department. During the combined inspection, inspectors with different backgrounds will look at the same corporation from various perspectives. A joint inspection is essential, because each inspector will have his own background, knowledge, and 'looking glass', with his or her mandate and checklist to see if the corporation complies with the rules. In the Netherlands, there is a shared uniform national inspection method that enables the different inspection agencies to run a Seveso II inspection. A joint inspection is also important for corporations: it results in fewer separate inspections.

At the end of a Seveso II inspection, inspectors will have an objective and reasoned opinion of the policy implemented by the corporation about major accidents and will have an opinion about the safety management system. In short, the inspection team checks to see if a corporation:

- has taken all measures necessary to prevent major accidents;
- has provided appropriate means for limiting the consequences;
- has informed the public;
- has data in the Safety Report that adequately reflects the real conditions:
  (a) demonstrating that a major-accident prevention policy and a safety management system for implementing it have been put into effect;
  (b) demonstrating that major-accident hazards have been identified and that the necessary measures have been taken to prevent such accidents and to limit their consequences for people and the environment;

(c) demonstrating that adequate safety and reliability have been incorporated into the design, construction, operation, and maintenance of any installation, storage facility, equipment, and infrastructure connected with its operation that are linked to major-accident hazards inside the corporation;
(d) demonstrating that internal emergency plans have been drawn up and supplying information to enable the external plan to be implemented in order to take the necessary measures in the event of a major accident;
(e) providing sufficient information to the competent authorities to enable decisions to be made in terms of the siting of new activities or developments around existing corporations (Art. 9 conjunction Art. 18, Seveso II Directive).

My present case study only focuses on the safety report requirements (Article 9) and the inspection (Article 18). Other obligations that fall under the Seveso II Directive – such as land-use planning legislation – are not included in this description of Seveso II enforcement. This chapter will take a closer look at only one of the agencies that is doing the Seveso inspections and the inspectors, the Environmental Protection Agency.

The inspectorate: DCMR Environmental Protection Agency

Our present research involves the most densely populated area of the Netherlands; it is called Rijnmond. The Dienst Centraal Milieubeheer Rijnmond (DCMR) Environmental Protection Agency is the regional environmental agency of the local and regional authorities operating in Rijnmond, the larger Port of Rotterdam area in the Netherlands. The DCMR was founded in 1972 in order to improve environmental protection in the Rotterdam-Rijnmond region. This region is a heavily industrialised area with refineries, waste incinerators, several waste dumping sites, many large chemical plants, metallurgy, food processing plants and about 19,500 smaller companies. Seveso inspections are applicable for about 140 corporations in the Rotterdam-Rijnmond area. The tasks of the DCMR include regulation of the industries and monitoring and assisting authorities on developing environmental policy.

The DCMR divides its tasks among different divisions: municipalities and small to medium-sized enterprises (SME), the Port of Rotterdam, and the industry and expertise centre. These divisions have subdivisions like 'monitoring and enforcement industry' or 'permits port and waste'. This study was limited to the two subdivisions that deal with Seveso inspections, because they mainly inspect chemical corporations that fall under the scope of Seveso. These subdivisions should inspect roughly 140 Seveso corporations every year. These corporations must take whatever measures that is necessary to prevent major accidents, to reduce consequences and need to have a high level of protection for people and the environment. They need to set up an internal emergency plan, a major accident-prevention policy, and implementation and provision of a safety management system. The safety management system consists of a fixed number of defined elements. The way in which these elements are implemented in a corporation depends on the hazards and risks that are present. In the next section, I discuss the practice of the inspectorate for the Seveso II Directive before discussing the method of research.

The practice of the inspectorate for the Seveso II Directive

Environmental inspectors of the DCMR are responsible for checking the compliance of upper and lower tier establishments. Since a Seveso inspection team consists of representatives of three enforcement agencies, inspectors participate in more than one team. Environmental inspectors do not have a routine office job, but divide their time between the office and outside visits. They are in charge of multiple Seveso corporations and therefore of inspecting premises, checking that improvements have been effected, and investigating incidents.

Seveso inspections in the Netherlands were being performed annually at corporations during the time of my research. Seveso corporations receive a letter in advance announcing these inspections; but for follow-up inspections, the corporations are not always informed in advance and could have surprise visits. Inspections are performed at Seveso corporations to report any violations of the Seveso II Directive. Inspectors check if the information in the safety report (if the corporation has one) conforms to the actual practice of the corporation.

During such inspections, the above-stated findings could lead to one of the four possible conclusions:

- First, it is determined that the situation complies with the regulations. In that case, there will be no follow-up activities.

- Second, the outcome is sufficiently severe to lead to a violation. In this situation, the corporation is explicitly responsible for carrying out improvements.
- Third, it is determined that the situation does not comply with the regulations and it can be regarded as a violation of the law. Within the inspection team, there will be mutual agreement on who will proceed with sanctioning. The corporation will be notified of this intention to sanction and after a period the violation is again controlled by the inspectors.
- Fourth and last, there is also a possibility that some findings individually are not sufficient enough to constitute an violation, but taken together they point to a structural deficit in an element of the safety management system and pose a violation.

The agencies inspect together, but they sanction according their own legislation, and currently it is not a joint activity. The Environmental Protection Agency sanctions according to the Environmental Management Act. Before discussing the data collection, I discuss the methods used in this study.

Method

Participant observations were made during the annual Seveso inspections and the follow-up at four chemical corporations in the Netherlands. These inspections took place between November 2010 and December 2011. The daily activities during these inspections are normally hidden from the general public, and we still know very little about how inspectors establish compliance (Hutter, 1997; Pautz, 2010). The observed inspections are the yearly inspections, as required by the Seveso II Directive (Article 18, Council Directive 96/82/EC).

Participant observation is done to gather data on the social interaction between environmental inspectors and chemical corporations. Confidentiality is paramount, since corporations that were inspected may have to deal with enforcement procedures. Since the objective of this research is to find out how inspectors make decisions, interpret the law, and apply the law, the 'atmosphere' in the interview room was important. The observations that were performed took place in areas where confidentiality was required to obtain information about among other things standard operating procedures and incidents. The distance between the subjects and the observer had to be small and I had to reveal my identity as an observer and therefore chose the role of observer as participant for the research. In this approach, it is possible that the presence of the researcher is revealed by his informants in the setting, but he doesn't really participate in the setting himself. The subjects under study are fully aware of the participation of the researcher. I was making notes but I did not participate in the inspection activities.

Box 8.1 Seveso inspections take place according to standardised phases*

1. *Preparation, individual inspector*: inspectors individually prepare their focus, subjects, and themes – for example, emergency procedures
2. *Preparation, inspection team*: the inspection team makes an inspection plan and schedules joint determination of inspection subjects, safety management elements, themes, and inspection goals

   Elements of a Safety Management System (SMS):
   SMS A. Major accident prevention policy
   SMS B. Employment and organisation
   SMS C. Identification and evaluation of major hazards
   SMS D. Operation control
   SMS E. Management of change
   SMS F. Planning for emergencies
   SMS G. Monitoring performance
   SMS H. Audit and review
3. *Carry out inspection*: kick off, walk all around the plant and corporation, do interviews, document review, and close out
4. *Conclusion of inspection*: close-out and inspection report. Inspection findings are categorised as:

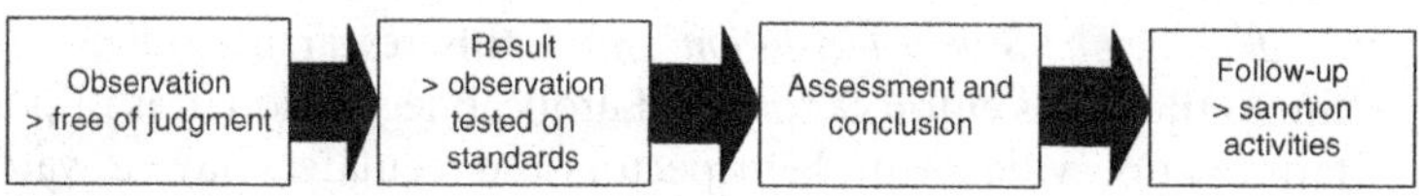

*This overview is a simplified version of the whole inspection process (LAT risicobeheersing bedrijven 2007).

We made observations of the interactions between the inspectors and regulatees during the daily routine of inspectors. We made full notes while accompanying inspectors on their visits. These notes contained explicit references to participants, interactions, routines, rituals, temporal elements, interpretations, roles, working environment, and behaviours. Data were not categorised while this exploratory phase of field work was still in progress. According to Hutter (1988: 210) one of the dangers of preclassification is that you can blind yourself to new information that it does not fall into an existing structure. The observed points of interest chosen in this research study are not only based on the literature, but also on an analysis of field notes after the phase of observing at four Seveso corporations. Although the list of points of interest will frame the analysis of the chemical corporations, I continued to make and record the broad, unfocused, and general observations to have a good base for future lines of research (see Hutter, 1988).

The research section of participant observation is divided into two main categories:

A. *Standardised phases of Seveso II inspections*: observations were made during the inspection process when inspectors were assessing a fixed number of points at each corporation (Mascini and van Wijk, 2008, 2009a, b).
B. *Points of interest*: research indicates that the following points of interest are relevant when doing participant observations of enforcement practices of regulation by field-level inspectors (Hutter, 1988, 1997; Mascini and van Wijk, 2008, 2009a, b). The points of interest in this research study are:

Inspectors' perception of their role in the inspection process. There is still very little written about environmental inspectors and the nature of their work. Since they have an important role in this inspection process, it is crucial to know what their perception of their role is. Besides that, the opinion of the inspectors of the regulated companies might influence their enforcement.

Attitudes towards Seveso legislation. Since this research studies how the Netherlands enforces specific European legislation, it is important to know what attitude inspectors and regulatees have towards this legislation. Versluis (2003) concludes that a positive opinion by inspectors of the legislation stimulates enforcement to actually take place, and if they consider it important, inspectors will put more time into enforcement. A negative opinion is likely to result in disinterest or hardly any enforcement. The opinion of the corporation might influence their compliance behaviour.

Proactive and reactive situations. Violations are made known to the Environmental Protection Agency in two ways: reported by a third party or discovered by an environmental inspector. In the document analysis and the participant observation component of this research study, both will be highlighted.

Working environment. This topic is not only referring to outside visits of inspectors while they are inspecting corporations. It refers as well to the working environment of the various inspectors at offices or at meetings with other inspection agencies.

Relationship with other agencies. Since Seveso inspections are carried out by three different inspectorates – the Environmental Protection Agency, the Occupational Safety and Health Inspection Agency, and the Fire Department – this was another central point in the observations.

Attitude towards violators. Since participant observations are done in the setting of an inspection, this research will describe the attitude of inspectors towards violators of the Seveso Directive.

Type of violation. A sanctioning practice of legislation is an interesting subject when studying enforcement. Seveso legislation is a European directive and therefore even more specific than general legislation. In the Netherlands, three enforcement agencies inspect together, but as mentioned earlier, they sanction according their own legislation. What kind of environment violations are there and how do environment inspectors sanction them?

Attitude towards small and large corporations. The Seveso II Directive involves two thresholds, an upper and lower tier, as mentioned before. Both kinds of corporations are studied in this explorative part of the research. Koolhaas (1990) observed that size of the regulated corporation influences the style of an inspector. He states that large corporations are complex and that inspectors are dependent on the good will of a corporation to cooperate. One of the questions that was posed by Versluis (2003) is investigated in this research, namely: To what extent do inspectors consider corporation size in their sanctioning practice?

Negotiations. According to Manning (1988), compliance is 'the process of extended and endless negotiation'. Negotiations have been identified by numerous studies (Carson, 1970; Cline, 2010; Hawkins, 1984; Hutter, 1988; McAllister, 2009; Shover et al., 1986) as a unique feature of regulatory enforcement. In the Netherlands, negotiations could be even more of an issue, since commission 'Oosting' concluded that inspectors should act more as enforcers and the policy of turning a blind eye during negotiations with regulatees should be abandoned immediately (Commissie-Oosting, 2001). To what extend are negotiations between inspectors and Seveso corporations of influence on compliance?

Data collection

The primary data for this research was collected between November 2010 and December 2011. The data consists of participant observations, interviews, and document analysis. The research started in September 2010 with the selection of a sample of chemical corporations within the 140 Seveso corporations. The objective of this selection was that the sample needed to be representative of the variety of work undertaken by Seveso II inspectors and of all the Seveso II corporations in this area. Since the intention was to observe whole inspection processes, this selection process has practical choices.

A Seveso inspection starts with a preliminary meeting among the inspectors of the Environmental Protection Agency, the Occupational Safety and Health Inspection Agency, and the Fire Department to determine the inspection subjects. Since this meeting forms the basis of the whole inspection process, I needed to observe these meetings too. A second choice was made on the basis of practicality. For preparation, the Seveso II inspection team needs a half day, the inspection itself could have a duration of seven days, and, after the inspection the team needed a half day as well. I decided to observe one Seveso II inspection each month. On that basis, the sample exists of 15 Seveso II corporations, and this chapter describes results of 4 of these establishments; this is the explorative part of the complete research study.

It took approximately ten months to get permission from the inspection agencies and permission of the corporations, for the inspection to be observed and for me to be present during the inspections. In addition to that, I needed to have personal protective equipment to enter the chemical corporations, a personal safety logbook, and obligatory safety training and exams. After the training and exams, I had more knowledge than before about the processes and did a small part of the safety training of employees of chemical establishments. Of course, I needed to know as well what to do when a sulphide chemical would accidentally be released during an inspection. All corporations and inspectors cooperated with the research.

Data analysis

Registered violations and enforcement activities of four Seveso corporations

Table 8.1 is an overview of the corporations that are under study in this research. In the table some key features are outlined, such as the nature of the process and the number of employees. As Table 8.1 shows, the processes in the corporations are different, and there are differences in the ages of the corporations and the number of employees. Since Seveso II has two thresholds, it is a normal characterisation of these

Table 8.1 Overview of four Seveso corporations

Corporation C01 process: filling and storage of propane, and carbon dioxide and storage of industrial gases in cylinders age: less than 10 years number of employees: 6 Seveso: lower tier	Corporation C02 process: transportation, warehousing, and handling containers with hazardous substances age: more than 50 years number of employees: 100 Seveso: lower tier
Corporation C03 process: producer of industrial gases of hydrogen and carbon monoxide age: more than 10 years number of employees: 50 Seveso: lower tier	Corporation C04 process: refinery age: more than 50 years number of employees: 570 Seveso: upper tier

laws and regulations. As mentioned earlier, the distinction between lower tier and upper tier is determined based on the permitted quantities of dangerous substances.

The database and archive of the Environmental Protection Agency were used for the document analysis. I selected all the reports for the four corporations that were available since the beginning of the Seveso Act inspections in the Netherlands (1999). Since knowledge on the database and archive was limited I started checking every report. Throughout this process I learned that environmental inspectors made a decision if an incident was an violation or not. Since this research only concentrates on violations of environmental regulations of chemical corporations, I decided to look exclusively at violations listed in these resources. From some corporations, I could find reports from violations and follow-up activities of notices, penalties, and lawsuits starting in 2000, but that was not true of all the corporations. I used all the available data on the violations recorded by the Environmental Protection Agency of the four corporations between 1999 and 2011 to get a complete overview of the environmental violations, which is presented below in Figure 8.1.

Figure 8.1 shows all the environmental violations the four corporations committed between 1999 and 2011. A total of 257 violations were committed in this time frame. The sources of these violations can be categorised as: complaints made by citizens or neighbouring corporations, the corporation itself (by submitting reports or making mentions

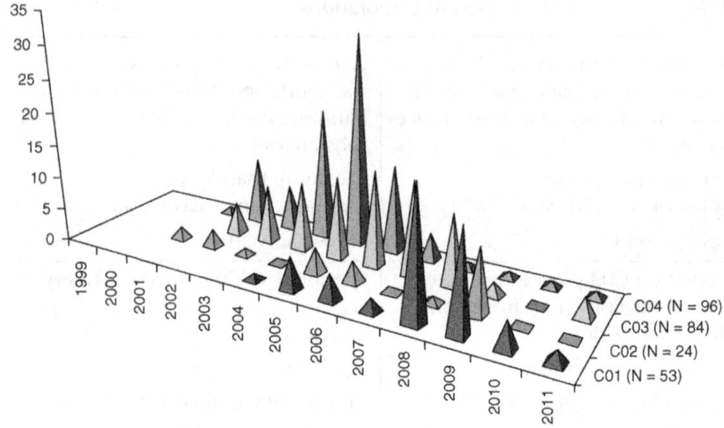

Figure 8.1 Total environmental violations of four Seveso corporations

of activities), the Environmental Protection Agency (by doing inspections or following up reports by corporations), and other causes. The backgrounds to these violations differ. Examples are: inappropriate storage of various hazardous substances and incidents and violations that were found on regular inspections re certain requirements or regulations. By taking a closer look at 2007 and 2010, we can note that the total number of violations was lower than in other years. This research needs to have more information in order to draw conclusions on this change in number of violations.

Table 8.2 reflects the enforcement actions of the Environmental Protection Agency on these violations. A total of 31 per cent of all the violations are enforced. An eye-catching observation is the low percentage (26%) of enforcement for Corporation C04, which has the highest amount of violations (N = 96). This low percentage can be explained by looking at the number of complaints; for this corporation, the number of complaints about odour and noise is higher than for the other corporations. If a corporation was dealing with these issues and implementing a solution for the problems that cause the complaints, the Environmental Protection Agency did not enforce it. In this explorative part of the study, I do not have a Seveso corporation of the same size to make a good comparison with Corporation C04. This study needs to do more research to draw more robust conclusions on this.

When I studied the enforcement activities of the four corporations, I found out that not all of them are related to the environment violations

Table 8.2 Enforcement overview

Year	Corporation C01	Corporation C02	Corporation C03	Corporation C04
1999		1		
2000		2		
2001		2		
2002		2	1	1
2003		1	2	0
2004	0	0	4	5
2005	1	0	1	3
2006	2	2	2	3
2007	1	1	3	4
2008	6	2	1	4
2009	8	1	3	1
2010	1	0	0	2
2011	1	1	2	3
Total	20	15	19	26

Table 8.3 Enforcement specifics

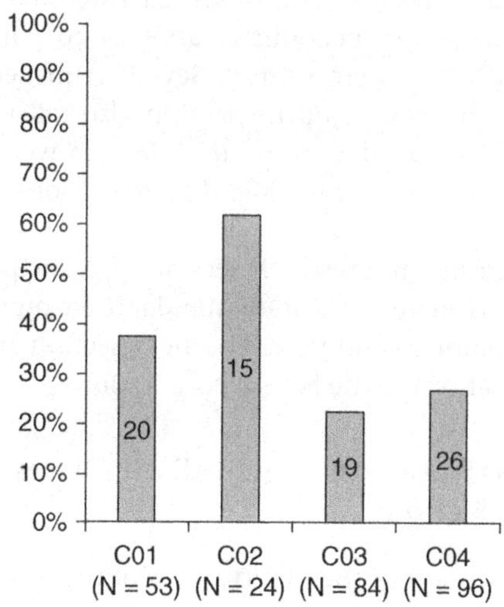

in Figure 8.1. This was an unsuspected finding. Table 8.3 shows that Corporation C02 has a total of 24 violations and 63 per cent of them are enforced (and that are 15 enforcement actions). Of those 15 enforcement actions, 9 enforcement activities are not related to the violations mentioned in Figure 8.1. I studied all of the enforcement activities that were not related to listed violations and found out that there are three immediate causes for this. The causes are: (1) reporting from the corporation itself, (2) an incident, and (3) during an inspection there was an observation, and at the moment it was not a violation. In the end all of these causes lead to enforcement activities.

Since none the above tables gives specific details on practice or details of the interactions between inspectors and regulatees, participant observation is the other method of research that was used in this study. The next section will describe the findings of participant observers at four Seveso corporations.

Participant observation at four Seveso corporations

I made observations during the daily routine of inspectors. I accompanied inspection teams at the preparation of an inspection, during an inspection, and after a Seveso II inspection. I observed a total of 14 inspectors and eight inspections at these four corporations. I observed for a total of 26 days at these eight inspections and saw seven different teams of inspectors and observed three inspections at one corporation and at two inspections at another corporation. Of these eight inspections, seven were normal Seveso II inspections and one was a follow-up. This follow-up inspection also fell under the scope of Seveso; it was done to determine if violations were ended. Of the 14 inspectors, 4 of them were female; this branch of industry is dominated by males.

As mentioned earlier, the research section of participant observation is divided into two main categories: standardised phases of Seveso II inspections and points of interest. The next section will describe the participant observations at the Seveso corporations.

Participant observation of standardised phases of Seveso II inspections

First, I give a description of the observations of the standardised phases of Seveso II inspections (see Box 8.1).

Observations of preparations of inspections

- During preparation for the inspection, inspectors decide which elements of the safety management system they are going to inspect. They decide that on the basis of what already had been inspected, so previous inspections, possible incidents, and past reports of inspections are needed. The whole inspection team decides which persons they would like to interview (by functions) during the inspection of the different elements and chooses documents that need to be sent by the corporation in advance. An average of six elements was discussed during one Seveso inspection. In the joint inspection system, they have examples of control lists for each element, but inspectors make audit trails themselves as well (different observations between 2010–2011).

- If inspectors have decided which safety management elements and other topics they are going to inspect, they discuss each subject in detail and apply it to the specific installations the corporation has. They discussed, for example, C02 – which scenarios they are going to inspect and how (observed January 2010).

Observations during walks around the plants and corporations

- Almost every inspection I observed started after the introduction with a walk around the plant. An inspection team is able to look into more details at small corporations such as C01 and C03. They can inspect details such as filling points, [and] storage of acetylene and cylinders. During the walk-around of the inspection at corporation C01, the inspection team notices that the employee who is working in the zone with risk of explosion carries a mobile phone that is turned on. He explains that he carries it to improve the corporation's contactability, although it is not allowed (observed November 2010).
- At large corporations, this part is more difficult and inspection teams decide in advance what installation will be more thoroughly inspected, although they see the whole site, but most of the time not by walking but cycling. The walk-around at corporation C04 is a cycle tour around the plant. Employees of the corporation show us the highlights. We received a detailed description of each production unit and we could ask or discuss something in great detail. During this inspection we did another field check on the topic of temporary changes of the emergency equipment. Since they were randomly picked, the corporation did not know this in advance and the inspection team has a more suprise visit of these parts of the plant (observed January 2011).
- These walks around the premises of a Seveso corporation give a good impression of the housekeeping and how these corporations work. Since corporations know that inspectors want to inspect in the field, it is of course possible that they clean up in advance (different observations between 2010–2011)

Observations of fixed elements of safety management systems (SMS)

- One corporation needed to show, for example, for safety management element B (employment and organisation), if they gave instructions to their employees and what kind of training they gave. Of course, staff needs to have the right knowledge to perform their duties within the corporation. At one of the corporations, safety instructions were not clear and it was also not clear what each employee had done for his training. The inspectors asked employees and [asked for] documentation from the corporation to check this topic. This safety management element needs to be updated by the corporation, since certain trainings have obligatory certificates and if it is not there, the employee cannot perform his task (observed November 2010).
- Some corporations have procedures that are applied worldwide and are difficult to bring in line with Dutch rules and regulations (observed December 2010).
- Inspection teams generally first start with the procedure of safety management element D (operation control), to check if it is updated to the current situation of the corporation, [and] if it is clear and in compliance with the Seveso II Directive. Corporations are obligated to identify major hazards (SMS element C), and therefore they could do safety reviews like FMECA (failure mode, effects and criticality analysis) and the result of the reviews [would] be certain measures. A question of an inspector could be: Is there a maintenance management system in place? (observed March 2011)
- Inspectors sometimes select an installation or a certain instrument and go through the whole safety management system to find out if this particular object is maintained, employees are certified and educated to work with it, if tests have been done regularly and [if] adequate technical measures are in place (and working) to prevent major accidents (observed January 2011).
- I observed different ways of inspecting the various safety management elements at these four corporations. An example is safety management element E, management of change, where the Fire Brigade Department made a special theme around this element and inspected this at all Seveso inspections that year in the Rotterdam-Rijnmond area. They made a special audit trail to check how corporations deal with temporary fire extinguishers and fire alarm systems. In this particular example, it involves means that are temporarily shut down, but are in a normal situation available. How do corporations

communicate that within their corporation, what do they put in place, and what are the consequences for operating the installations? (observed March 2011)

- In other inspections, corporations needed to show the inspectors how they approach this and use it within their corporation. Larger corporations have whole systems for this, since changes could happen every day, for example changing of a pump, and more people need to approve this change. Sometimes they first needed to do a safety review, before they decide to change their operation (observed January 2011)
- A question an inspector could ask for safety management element G, monitoring performance is: Are there procedures for an ongoing systematic assessment of safety performance in relation to the objectives of the policy to prevent major accidents? The smaller corporations in this explorative part have difficulties in dealing with this element and both had violations. The larger ones have whole systems for it to keep track of their output, especially for safety performance indicators (different observations between 2010–2011).

Observations of conclusions of inspections

- *After the inspection days at the corporation, the inspection team discusses the results among themselves. They prepare a presentation of the main findings and present the results of the inspection to the corporation. At corporation C01, the inspectors discussed the main findings just after the last inspection subject without the presence of the corporation. After that they presented their main findings on the same day (observed November 2009).*
- At all the other inspections, the inspectors have a meeting on another day in advance of the presentation at the corporation. At those meetings they discuss the inspection results and make a presentation together. Then there is room for discussion on the inspection results and time to talk details thorough (different observations between 2010–2011).
- After that the team of inspectors must draw its conclusions in an end report. In three cases of these eight inspections, the inspectors needed to resort to enforcement activities because of violations of Seveso regulations. They sometimes made these decisions together and sometimes they do the reinspections together as well.
- The offenses were on the following elements: organisational and personnel (B), identification and evaluation of major hazards (C), operational control (D), planning for emergencies (F), monitoring performance (G) (different observations between 2010–2011).

Participant observation of points of interest

The second main category of the research component of participant observation is 'points of interest,' which are described below.

Inspectors' perception of their role in the inspection process

- Hutter says that inspectors of law enforcement agencies have a variety of tactics and enforcement devices at their disposal, ranging from informal, conciliatory techniques to the formal and more coercive tools (1988). During the eight inspections I observed at the four corporations, I saw multiple styles, even among inspectors of the same inspection team.
- For example, at corporation C0,1 the inspection leader had a formal and more strict role. He inspected safety management element B, employment and organisation, and after a few questions it was clear the corporation did not have any knowledge of these specific regulations. The inspectors had two roles at that inspection: they clearly explained what the corporation did wrong and how they could improve the mistakes and helped them comply with the rules, but they were also checking the 'old violations' and behaving more like 'police officers'. In this specific situation, the 'police-officer' role was the inspection leader of the inspection team. The violations of the corporation continued for a longer period (observed November 2009).

Attitudes of inspectors towards legislation

- When I accompanied inspectors during their daily activities, I drove with them to the corporations and we had conversations on this topic. Some inspectors say that the Seveso Directive has too many rules and that they are not clear. Other inspectors have a different opinion and regard the Seveso Directive as a flexible and therefore [a] good, applicable directive, since small and large corporations need to abide it.

Proactive and reactive situations

- Both situations appear in this research study. During the walk around corporation C04, the inspection team observed that pipes were hanging loose at an installation. During this outside visit, the inspection team was accompanied by employees of the corporation, and they noticed it as well. After the walk-around, inspectors and employees of the corporation discussed this observation.

Corporation C04 took this observation seriously and immediately afterwards they sent a team over there to check it in more detail. They find out that the pipes are in use and contain hydrogen. During this inspection period, the corporation reported that the problem is solved. The inspectors checked it with a field check and it was correct. The violation did not show up in the final inspection report (observed January 2011).

- There are two ways [kinds] of reactive situations: first, in response to complaints and secondly, in response to accidents and incidents. An example is: A police officer of the Harbour Police reports to the Environmental Protection Agency that corporation C01 is doing business activities and he is alarmed by the fact that the corporation has dangerous goods. This report is the immediate cause for an environmental inspector to perform an inspection, since they were not aware of the existence of the corporation before that.

Working environment

- Most of the time these corporations are located in an industrial area and not close by neighbourhoods. Some corporations have a barrier in front of their premises and at others you can easily walk in. All four corporations had some kind of a registration system for visitors (different observations between 2010–2011).
- In advance of observing these inspections, I needed to do an obligatory safety training, which is combined with a safety passport. In this exploratory stage, only one corporation asked me to show it [the passport] to them. Of course this is probably related to the fact that I was always accompanied by inspectors, but it was strange that not all corporations demanded me to show it to them. Especially since dangerous goods at all corporations were stored, produced, or handled, and in full production while I was there (different observations between 2010–2011).

Relationship with other agencies

- Since Seveso inspections are a joint effort in the Netherlands, inspectors need to work together during the whole process of an inspection. Some inspection teams consisted of two inspectors, but in most inspections I observed, all three representatives of the agencies were involved. If there was any discussion among them, that always took place at the meeting after the inspection days and not in front of the corporation [people]. These discussions were of course related to the findings during the inspection, whether they were violations or not (different observations, between 2010–2011).

Negotiations

- Large corporations can and do use their power to manipulate the labelling process and minimise sanctions (Hutter, 1997). The earlier mentioned violation of corporation C04, which did not end up in the inspection report, is a good example of negotiations. On the other hand, the inspector told me that the pipe contained a less dangerous product and he assessed the risk as low. This relates to the fact of practicability, where inspectors are willing to accept anything if the necessary results are achieved.

Comparison of document analysis and participant observation

As already mentioned in this chapter, two data sources were used for this research study: document analysis of registered violations and enforcement activities, and participant observation of four Seveso corporations. This section shows the importance of both data sources in this research by describing two examples. In the document analysis, there was a report of an violation of corporation C01 in 2005. This violation was noticed during an inspection; it involved the violation of using a mobile phone in an explosive area, where it was not allowed. During the inspection of November 2010, inspectors interviewed an operator while he was working at an installation. They asked him if he carried a mobile phone and if it is currently working, and they found out that the mobile phone is turned on. This is a violation of the corporation's own rules and regulations and also a violation of the Environmental Management Act. The inspectors decided that the violation would not receive a follow-up or enforcement. I wonder whether, if they had known about the history of the violation in 2005, which makes the violation in 2010 a case of recidivism, they would make the same decision not to sanction the violation. Without the participant observation of the inspection of corporation C04, I could not have known that the corporation had a violation that is described in the part on proactive situations in this chapter, since the violation did not show up in the inspection report in the end. These two examples show that both data sources are valuable when studying the implementation of field-level inspectors, especially when you do case study research on corporations. Since this chapter is the result of an explorative part of a complete research study, we need to observe more inspections to see whether some of these findings are unique or repetitive. Document analysis only shows the end result of these proactive and reactive enforcement activities; participant observation gives a more complete view what is really going in the field.

Conclusion

Seveso inspections in the Netherlands are the objects of this research study. These inspections have been performed for 10 years under the auspices of the Seveso Directive on Major Hazards within the European Union. In the Netherlands, the inspections are carried out annually by inspectors from the Environmental Protection Agency, the Occupational Safety and Health Inspection Agency, and the Fire Department. There is little specific study of the role of field-level inspectors in the implementation of environmental regulations. Our present research begins to respond to this gap by focusing on a small subsection of field-level inspectors in environmental policy, environmental inspectors in the Netherlands. The main purpose of this explorative part of the research is to obtain an understanding of environmental inspectors and their work, to collect more specific data about their working methods, and to establish whether participant observation served as a suitable method of investigation. Understanding these interactions have provided insights into how regulators approach their interactions with regulatees and how they ensure environmental compliance. The primary data for this research was collected between November 2010 and December 2011. The data consists of participant observations, interviews, and document analysis. Participant observations were made during the annual Seveso inspections and the follow-up at four chemical corporations in the Netherlands. These inspections took place between November 2010 and December 2011. During these inspections of the premises, inspectors got an impression of a part of a corporation. Since time is limited, they see a 'snapshot' of the corporation's condition, the abilities of the personnel, and gain an impression of what was really going on. But the visible presence of inspectors is thought to hold a deterrent value, and it shows that inspectors enforce the law. Since this chapter is an explorative part of our complete research study, we need to observe more inspections to see whether some of these findings are unique or repetitive. Document analysis only shows the end result of these proactive and reactive enforcement activities; participant observation gives a more complete view what is really going in the field.

Acknowledgements

I would like to thank the Environmental Protection Agency and their inspectors for their generous cooperation in this study. I thank Ben Ale, Wim Huisman, and Ellen Jagtman for their valuable comments on a previous version of this chapter.

Notes

1. Directive 87/216/EEC of March 1987. OJ L 085, 28/03/1987, pp. 0036–0039.
2. Directive 88/610/EEC of November 1988. OJ L 336, 07/12/1988, pp. 0014-0018.

References

Bardach, E. and Kagan, R. (1982) *Going by the Book: The Problem of Regulatory Unreasonableness*. Philadelphia: Temple University Press.

Besluit van 27 mei 1999 tot vaststelling van het Besluit risico's zware ongevallen 1999 en tot herziening van enkele andere besluiten in verband met de uitvoering van Richtlijn nr. 96/82/EG van de Raad van de Europese Unie van 9 december 1996 betreffende de beheersing van de gevaren van zware ongevallen waarbij gevaarlijke stoffen zijn betrokken, Staatsblad 1999, 234.

Braithwaite, J. (1984) *Corporate Crime in the Pharmaceutical Industry*. London: Routledge and Kegan Paul.

Braithwaite, J. (1985) *To Punish or Persuade*. Albany: State University of New York Press.

Carson, W. G. (1970) 'Some sociological aspects of strict liability and the enforcement of factory legislation', *Modern Law Review*, 33: 396–412.

Clinard, M. B. and Yeager, P. C. (1980) *Corporate Crime*. New York: Free Press.

Cline, K. (2010) Negotiating the superfund: are environmental protection agency regional officials willing to bargain with states? *The Social Science Journal*, 47: 106–120.

Commissie-Oosting (2001) *De vuurwerkramp*. Eindrapport van de Commissie Vuurwerkramp Enschede. Den Haag: Enschede.

Council Directive 96/82/EC on the control of major-accident hazards involving dangerous substances, OJ 1997 L 10/13.

Directive 87/216/EEC of 19 March 1987. OJ L 085, 28/03/1987, pp. 0036–0039.

Directive 88/610/EEC of 24 November 1988. OJ L 336, 07/12/1988, pp. 0014–0018.

Geis, G. (1967) 'The heavy electric equipment antitrust case of 1961', in M. B. Clinard and R. Quinney (eds) *Criminal Behaviour Systems: A Typology*. New York: Holt, Rinehart and Winston, pp. 139–150.

Hale, A., Hopkins, A. and Kirwan, B. (2002) *Changing Regulation: Controlling Risks in Society*. Oxford: Pergamon.

Hawkins, K. (1984) *Environment and Enforcement: Regulation and the Social Definition of Pollution*. Oxford: Clarendon Press.

Huisman, W. (2001) *Tussen winst en moraal. Achtergronden van regelnaleving en regelovertreding door ondernemingen*. Den Haag: Boom Juridische Uitgevers.

Hutter, B. M. (1988) *The Reasonable Arm of the Law*. New York: Clarendon Press.

Hutter, B. M. (1997) *Compliance: Regulation and Environment*. Oxford: Clarendon Press.

Kagan, R. A., Gunningham, N. and Thornton, D. (2011) 'Fear, duty, and regulatory compliance: lessons from three research projects', in C. Parker and V. L. Nielsen (eds) *Explaining Compliance Business Responses to Regulation*. Cheltenham, Northampton: Edward Elgar, pp. 37–58.

Koolhaas, E. (1990) 'Milieu, handhavingsstrategieën en beleid', *Justitiële Verkenningen*, 16: 92–109.

LAT risicobeheersing bedrijven (2007) *Werkwijzer BRZO 1999*. Available at http://www.latrb.nl/instrumenten-0/werkwijzer-brzo-ii-0/. Accessed 8 August 2012.

Lipsky, M. (1980) *Street-Level Bureaucracy: Dilemmas of the Individual in Public Services*. New York: Russell Sage Foundation.

Manning, P. K. (1988) 'Review of Braithwaite, J. To punish or persuade', *British Journal of Criminology*, 28(4): 559–561.

May, P. J. and Winter, S. (2000) 'Reconsidering styles of regulatory enforcement: patterns in Danish agro-environmental inspection', *Law and Policy*, 22: 143–173.

May, P. J. and Winter, S. (2011) 'Regulatory enforcement styles and compliance', in C. Parker and V. L. Nielsen (eds) *Explaining Compliance Business Responses to Regulation*. Cheltenham, Northampton: Edward Elgar, pp. 222–244.

May, P. J. and Wood, R. S. (2003) 'At the regulatory front lines: inspectors' enforcement styles and regulatory compliance', *Journal of Public Administration Research and Theory*, 13(2): 117–139.

Mascini, P. and van Wijk, E. (2008) '"Vis ruikt nou eenmaal zo". Responsive regulation door de Voedsel en Waren Autoriteit', *Tijdschrift voor Criminologie*, 50(2): 114–129.

Mascini, P. and van Wijk, E. (2009a) *Responsive regulation bij de Voedsel en Waren Autoriteit*. Den Haag: Boom Juridische uitgevers.

Mascini, P. and van Wijk, E. (2009b) 'Responsive regulation at the Dutch Food and Consumer Product Safety Authority: an empirical assessment of assumptions underlying the theory', *Regulation and Governance*, 3: 27–47.

McAllister, L. K. (2009) 'Dimensions of enforcement style: factoring in regulatory autonomy and capacity', *Law and Policy*, 32(1): 61–78.

McCaffrey, D. P., Smith, A. E. and Martinez-Moyano, I. J. (2007) 'Then let's have a dialogue': interdependence and negotiation in a cohesive regulatory system', *Journal of Public Administration Research and Theory*, 17: 307–334.

Pautz, M. C. (2010) 'Front-line regulators and their approach to environmental regulation in Southwest Ohio', *Review of Policy Research*, 27(6): 761–780.

Pearce, F. and Tombs, S. (1998) *Toxic Capitalism: Corporate Crime and the Chemical Industry*. Dartmouth: Ashgate.

Pearce, F. and Tombs, S. (2012) *Bhopal: Flowers at the Altar of Profit and Power*. Lincolnshire: CrimeTalk Books.

Shover, N., Clelland, D. and Lynxwiler, J. P. (1986) *Enforcement or Negotiation? Constructing a Regulatory Bureaucracy*. Albany: State University of New York Press.

Simpson, S. S. and Rorie, M. (2011) 'Motivating compliance: economic and material motives for compliance', in C. Parker and V. L. Nielsen (eds) *Explaining Compliance Business Responses to Regulation.* Cheltenham and Northampton: Edward Elgar, pp. 59–77.

Versluis, E. (2003) *Enforcement Matters*. Utrecht: Utrecht University.

9

The Uneven Geography of Environmental Enforcement INGOs

Paul B. Stretesky and Olga Knight

Introduction

Green criminology can be broadly defined as the study of crimes against the environment (South, 1998). A growing number of criminologists have become involved in green criminology to better understand the nature, causes, and consequences of environmental crime (Agnew, 2011; Brisman, 2008; Burns, Lynch and Stretesky, 2003, 2008; Eman, Meško and Fields, 2009; Gibbs et al., 2010; Lynch, 1990; Pellow, 2004; Situ, 1997; South, 1998; Stretesky, 2008; White 2008). Nevertheless, within this relatively new area of study, little attention has been devoted to role that international nongovernmental organisations (INGOs) play in environmental protection.[1] The neglect of environmental enforcement INGOs in the literature is surprising, because early green crime scholars have emphasised power, harm, and justice. Lynch (1990: 3), for instance, points out that 'powerful groups manipulate and use race, class, gender and the environment to preserve the basis of their power'.

As a result, the reaction to environmental harm and ecological disorganisation by civil society should be of considerable interest to green criminologists.[2] However, even while disciplines such as sociology and political science have accounted for the rise of civil society in studies of 'global environmental governance' and 'earth systems governance,' green criminologists have been slow to study the non-state transnational responses to environmental crime and deviance.

To address the neglect of INGOs in green criminology, we organise this chapter around three issues. First, we draw upon treadmill of production theory to describe how INGOs may advocate for environmental protection within the existing global political economy. We demonstrate that while the state is central to preventing ecological disorganisation, nongovernmental organisations may play an important role in this process. Second, we examine how environmental enforcement INGOs can work to influence levels of green crime that are of specific interest to green criminologists and state environmental enforcement agencies. Third, we draw upon the concept of environmental justice as developed within green criminology to examine the distribution of INGO headquarters across countries. Specifically, we seek to determine if environmental enforcement INGO headquarters are more likely to be located in high-income countries than in low-income countries. The location of environmental enforcement INGO headquarters with respect to income is a significant environmental justice issue.[3] First, INGO headquarters often influence where and when environmental enforcement–related activism and operations will take place. If headquarters are located in high-income countries, then organizations may not adequately reflect the concerns of low-income countries (Gómez, 2008). Second, if environmental enforcement INGOs tend to be concentrated in high-income countries it could have a negative impact on the development of an environmental enforcement civil society in low-income countries (Gómez, 2008; Smith and Wiest, 2005). Moreover, the agglomeration of INGO headquarters in high-income countries is described as potentially impacting low-income countries in a neocolonialist and unjust fashion (Gómez, 2008). An uneven geography of INGOs across the globe may detract from a global civil society that is engaged in environmental protection by allowing ecological disorganisation to accelerate as treadmill policies continue. As far as we are aware, these issues have yet to be taken up in any study of ecological disorganisation.

INGOs, crime, and criminal justice in the treadmill of production

International organisations operate and advocate for environmental protection within a political economy of production. To explain how INGOs may fit into this political economy and act to reduce ecological disorganisation we draw upon the *treadmill of production theory* (Schnaiberg, 1980). Treadmill of production theory was developed by Alan Schnaiberg to explain the increasing levels of environmental harm that he observed after World War II (see also Gould, Pellow and Schnaiberg, 2008). The engine that drives the treadmill of production is capitalism and its expansion. As production increases and becomes more efficient, the treadmill accelerates and workers run to keep up with increasing production; at the same time the treadmill casts them off as the unemployed and underemployed. In addition to these class-based concerns, increases in production lead to significant exploitation of the environment (Lynch, 1990). Specifically, as production increases, natural resources are needed to fuel the treadmill and the treadmill releases production-related pollution (Gould et al., 2008).

Treadmill of production theory suggests that most people falsely believe that society can only advance if the economy grows. This faith in economic expansion means that various actors in the system work to ensure fiscal growth. According to Schnaiberg (1980) the problem of ecological disorganisation became apparent with the rise of the chemical revolution and the increasing use of nonorganic chemicals in the production process. As Greenaway (1992: xii) noted in the foreword to the reprinted edition of *The Chemical Revolution*, 'Chemistry has generated the world's dominant industry'. These chemicals were used to expand production and accelerate the treadmill (Schnaiberg, 1980).

Although the chemical revolution increased production, it also caused ecological disorganisation. First, to facilitate production increases, a greater level of natural resources must be withdrawn from the environment. Schnaiberg (1980) labelled the extraction of natural resources 'ecological withdrawals' and noted that they often cause significant disruption to ecosystems. For example, consider that coal is a natural resource that is needed to create the energy used to manufacture the products sold in the marketplace. Between the years 1981 and 2010, worldwide coal production increased from 3,836.1 million tonnes to 7,273.3 million tonnes or 4.23 million tons to 8.02 million tons (British Petroleum Inc., 2011). The extraction of coal from the environment through surface mining, for example, is associated with the use and release of a significant amount of chemicals into the environment. According to Palmer et al. (2010: 149) the 30-year increase in mountaintop removal to keep up with needed energy demands is highly destructive to the environment, and 'contaminants persist in streams well below valley fills, forests are destroyed, headwater streams are lost, and bio-diversity is reduced'. Surface mining is a global concern and it is the preferred method of coal extraction. Kobayashi (2009) notes that coal production is particularly harmful in low-income countries where regulations are weak, including Bangladesh, where foreign companies are extracting significant amounts of coal.

In many countries such as Bangladesh, multinational corporations are able to remove natural resources from the environment while causing considerable environmental harm. For instance, in Bangladesh coal resides in swampy areas that are home to mangrove forests. Mangroves are important to the ecosystem and filter out pollutants, prevent flooding, and sequester carbon. However, open-pit coal mines threaten to destroy the water quality and function of Bangladesh's mangrove forests. Damage to these resources harms the local populations, since many people rely on the mangroves for their livelihoods. Thus, coal surface mining occurs despite the protests of the citizens who rely on the mangroves (Kobayashi, 2009). The lack of regulations and oversight has left many scientists calling for better regulation of strip mining across the globe (Palmer et al., 2010).

Increases in production also result in increases in the release of toxic chemicals into the environment in the form of air, land, and water pollution.[4] Schnaiberg (1980) labelled these releases 'ecological additions' and noted that they can disrupt the environment and cause significant harm to humans and nonhumans. For example, the American Lung Association (2010) notes that 1 in 10 residents in the United States is chronically exposed to harmful particulate matter. In addition, the World Health Organization (2011) reports that 1.3 million deaths occur annually across the globe because of air pollution.

Perhaps the most noted example of ecological additions in the form of air pollution occurs in China, where coal-burning power plants release considerable pollution. High levels of coal consumption that produce this air pollution are driven by the need for energy that is used in the production and manufacture of consumer goods. In 2005, for instance, the mean concentration of large particulate matter in the air (that is, particles 10 micrometers in diameter) exceeded the World Health Organization's annual guideline value of 20 $\mu g/m^3$ by over five times (109 $\mu g/m^3$). Particulates in the air are well known to cause respiratory and cardiovascular diseases and result in a large number of premature deaths (Samet et al., 2000). In a recent study, Matus et al. (2012) estimate the economic costs to human health associated with the generation of harmful particulate matter in China alone are approximately US $64 billion, or nearly 9 per cent of the Chinese gross domestic product. In the end, ecological additions and withdrawals increase because the investments in production-enhancing toxic technology were paid with the future profits that are to be derived from increased production. Thus, the treadmill of production represents the continual and accelerated drive to expand production at the cost of ecological disorganisation.

There are three important sets of actors in the treadmill of production, each of which has an incentive to keep the treadmill running. First, there are those firms that focus on increasing production for profit. These companies are driven by the bottom line and promote production at the cost of environmental destruction. In many cases international corporations may locate in countries with the lowest wages and weakest environmental laws in order to minimise the costs associated with the extraction of natural resources and production.

Labour is the second set of actors in the treadmill of production theory. Labour may be motivated to support increases in production to promote the creation of jobs that helps increase wages and reduce unemployment. In short, the implementation of technology and production often comes with the promise of financial investment in areas that are economically depressed (Gould et al. 2008). Thus, labour unions may often support policies that increase production and therefore cause more ecological disorganisation because they believe these policies are beneficial to workers. In reality, however, this benefit to workers is not realised, as technology generally displaces workers (Gould et al. 2008). For example, Stretesky and Lynch (2011) found that increases in mountaintop removal over traditional forms of underground mining are associated with a significant decrease in employment over time, because surface mining relies on fewer workers, as excavation is done with machines and chemical explosives.

Third, state actors also often support the expansion of production because it brings additional revenue in the form of taxes. Taxes legitimate the role of the state and can be redistributed to achieve policies consistent with liberal economics (Schnaiberg, 1980). Thus, the state may also support the development of businesses that may harm the environment. In some cases, states may even lower their environmental standards to attract business. For example, Woods (2006: 174) discovered that 'political officials may be motivated to reduce regulatory stringency to gain a competitive advantage over their neighbours, thereby creating an aggregate movement toward the lowest common denominator.'

This chapter investigates a fourth actor, INGOs. Specifically, we examine the importance of INGOs and their ability to pressure the state to reduce ecological disorganisation. According to Schnaiberg (1980), the state is an arena of contention where corporate and environmental interests collide. Thus, the state has the potential to impact the ecosystem in opposite ways. On the one hand, corporations put pressure on the state to engage in the 'race toward the bottom'. On the other hand, environmental organisations, such as INGOs, are part of a civil society that can pressure states to demand that firms reduce emissions, thus slowing the treadmill. In short, the state may help manage conflicts concerning the disorganising effects of production on the environment (Obach, 2004). The idea that organisations can pressure the state is of central importance within treadmill of production theory. We are interested in those INGOs that are part of that civil society that influences environmental enforcement. In particular we will examine where these organisations are headquartered relative to economic conditions to determine if there is potential for global inequality in this new form of emerging civil society.

INGOs represent an increasingly important external force on state actors (Boli and Thomas, 1999), both challenging and aiding states. In short, INGOs serve actively as environmental police and prosecutors and advocate for better environmental law and stronger enforcement. Schnaiberg (1980) has argued that the state will not place constraints on corporate actors without considerable public pressure. Thus, Schnaiberg's message regarding community organisations is somewhat mixed. He notes that while citizen groups have not yet forced states to adopt steady-state policies, these organisations remain the only hope for challenging ecological disorganisation. Nevertheless, there is some anecdotal evidence that citizen organisations have helped to reduce the negative externalities associated with production though their direct involvement in environmental problems and through pressure on state enforcement. We investigate the distribution of two types of environmental enforcement INGOs.

Types of environmental enforcement INGOs

In general, INGOs engage in advocacy and operations work. The INGOs that we examine are nonprofit groups that are independent from the government and advocate for stronger environmental laws or environmental enforcement. The international focus of INGOs means that although they are headquartered in one country, the organisations advocate in several countries. INGOs may lobby, denounce, and even influence states to act on environmental issues. In addition to advocating for stronger environmental policy and practices regarding environmental crime, INGOs may also take on an operations function in order to enforce or prosecute environmental crime. For example, the Sea Shepherd Conservation Society engages in operations and 'confrontational tactics due to its sophisticated use of international law, and by taking advantage of overlapping international legal regimes' in order to prevent wildlife violations (Bondaroff, 2011). Moreover, environmental organisations are able to operate in ways that monitor environmental conditions so that they can shift the 'balance of power between activists, state regulators, and private firms based on their ability to contest official accounts of environmental quality' with information on environmental performance (Overdevest and Mayer, 2008: 1497).

In reality, the distinction between advocacy and operations INGOs is not so clear, and many groups may engage in both types of actions. For example, Gould et al. (2008: 104) have argued that challenges to the treadmill of production occur when environmental organisations 'disrupt, monitor, and shame transnational corporations into behaving responsibly'. Thus, organisations shape formal governmental social control of environmental harms through a variety of methods that include protest, political pressure and campaigns, campaign contributions and political support, petitions and lobbying, and actual enforcement-related efforts with state support (Burns et al., 2008). Thus, while Schnaiberg (1980) clearly notes that capital is the dominant actor that drives ecological disorganisation, it is also clear that citizen groups have the ability to influence state actors in the environmental crime arena.

Advocacy INGOs

INGOs may serve as advocacy groups. That is, these organisations may advocate for stronger environmental laws or for stronger enforcement of environmental laws. For example, the Coalition for a Clean Baltic (see http://www.ccb.se/about.html) is an INGO that 'promotes the protection and improvement of the Baltic Sea environment and natural resources...at the international and national policy levels'. Its website notes that the group approaches this goal primarily through lobbying, which includes creating 'public opinion about Baltic Sea issues', and through joint appeals to organisations such as the Baltic Marine Environmental Protection Commission, which creates policy for the region.

In terms of advocacy, INGOs have gained considerable legitimacy in the international arena with respect to environmental protection. The legitimacy of INGOs is bolstered by the fact that no state has authority over any other. Thus, INGOs serve an important global function because they can direct the global culture with respect to important collective values. For example, Boli and Thomas (1997: 181) suggest that INGOs cannot dominate in the conventional sense. INGOs have little sanctioning power, yet they act as if they were authorised in the strongest possible terms. They make rules and expect them to be followed; they plead their views with states or transnational corporations, and they express moral condemnation when their pleas go unheeded.

Thus, INGOs that advocate for environmental law can influence states because they can negotiate issues of sovereignty more easily than states can. Boli and Thomas (1997) point out that INGOs are in a position to influence states about how they should behave when they are not engaging in environmental enforcement. Thus, INGOs may help shape the environmental crime agendas and state behaviour toward environmental crime. Frank (1997) also suggests that INGOs are significant actors in environmental sector. He notes that these organisations have had an enormous impact on the way states developed their environmental policy because of their impact on the targets of state authority. In short, INGOs may stigmatise violators and direct attention to corporate environmental destruction (Boli and Thomas, 1997).

Advocacy INGOs may also be able to influence the state by pushing for greater levels of enforcement. For example, Stretesky et al. (2011) discovered that environmental organisations play an important role in the level of environmental enforcement in Florida (US). Similar arguments have been made internationally, where citizen activism around environmental issues has emerged as a response to global ecological troubles. As Gould, Pellow, and Schnaiberg (2008: 101) observe:

> Transnational social movement organizations are proliferating and have been since the dawn of the post-World War II era.... [Social movement organizations] are now widely acknowledged as formidable players in international politics because they are creating new global norms and practices among states, international bodies, and corporations, and transforming new ones. Such non-state actors can be viewed as sources of resistance to globalization from below.

To be sure, INGOs that operate within the environmental arena have been having an impact on the shape and focus of environmental crime across the globe.

Operations INGOs

Environmental enforcement INGOs also engage in operations-related activities such as training law enforcement, monitoring for violations, and even prosecuting environmental offenders. In some cases, governments may even deliver enforcement programs through INGOs that operate in their countries (Srivastava, Schwartz and Austin, 2012). These organisations usually work within the confines of existing laws to enhance environmental enforcement and to aid law enforcement. In some cases, however, these types of INGOs may enforce laws, through direct action – for instance, when law enforcement is not concerned with enforcement (Bondaroff, 2011). Operation INGOs may work in collaborative community partnerships that may involve citizen advocacy groups (Skogan and Hartnett, 1997). O'Rourke and Macy (2003) point out that the focus on civic engagement to deal with an environmental monitoring crisis is similar to the traditional policing crisis that led to community-oriented policing ideals. For instance, O'Rourke and Macey (2003: 383) observe that 'public participation in environmental issues is supported for its potential to provide additional (often low-cost) sources of information to government agencies, increase acceptance of and confidence in government decisions, educate and empower community members on issues that affect them, and advance democratic ideals.'

Overdevest and Mayer (2008: 1497) suggests that the development of environmental monitoring organisations is not surprising because 'local antitoxic organizations have realised the importance of collecting and diffusing information in order to influence industrial firms' environmental performance'.

For example, some INGOs engage in monitoring practices by taking air samples in what have been described as 'bucket brigades'. Global Community Monitor (http://gcmonitor.org/index.php) is one bucket-brigade organisation that operates in over a dozen countries (for example, Australia, the United States, Zambia, Nigeria, South Africa, Thailand, Israel, the Philippines, India, Kazakhstan, Netherlands, Ireland, England, Spain, Barbados, and Curacao) and trains community organisations in how to monitor air released from industrial facilities. Information obtained from monitoring can help various organisations' direct enforcement and can influence advocacy efforts. Bucket brigades and other types of international monitoring organisations have developed as citizen-led community policing groups that help identify violations when formal law-enforcement organisations lack resources to do so (O'Rourke and Macey, 2002). These INGOs resemble citizen-watch programs that have developed as part of the community-oriented policing efforts that emerged in many developed countries during the 1990s (Oliver, 1998). In effect, these organisations supplement and help direct the ability of authorities to 'police' the environment and be responsive to community needs (Lynch and Stretesky, 2011).

In other direct types of enforcement operations, INGOs may collect information and use it in the courts to push for compensation and punishment. For example, Earthjustice (http://earthjustice.org/about) describes itself as an organisation that 'works through the courts on behalf of citizen groups, scientists, and other parties to ensure government agencies and private interests follow the law'. Like many environmental organisations, Earthjustice also engages in lobbying-related activity. For instance, one attorney for Earthjustice suggests:

> In public interest work, winning cases is often only half the battle. To keep our victories from being undone by legislative action, I often testify before the Hawaii Legislature or work with Earthjustice's policy folks in Washington, DC to influence lawmaking at the national level. I draft press releases, fact sheets, op-eds and other materials, hold press conferences, and speak at conferences to educate the press and public about the importance of our litigation efforts and to remind them about the central role environmental protection plays in improving the quality of life in Hawaii. (cited in Yale Law School, 2011: 19)

Some organisations also fight environmental crime and deviance with illegal tactics that are labelled as 'ecotage,' or 'monkey-wrenching' (Gottchalk, 1998). These more radical, direct-action tactics are associated with groups such as Earth First! that may not be officially recognised as INGOs and often carry out direct action that violates the law (Vanderheiden, 2005). For instance, Earth Liberation Front uses arson to stop environmental damage. One of its initial members, Rod Coronado, explains that he once used illegal tactics because 'after years of rescuing animals from laboratories, it was heartbreaking to see those buildings and those cages refilled within the following days. And for that reason, arson has become a necessary tool' (cited in Schorn, 2005).

There is little doubt that environmental organisations have often opposed the state and corporations through various methods and operations, but have also helped to direct the state's attention to environmental problems and issues by threatening state legitimacy. Dorn, Van

Daele, and Vander Beken (2007: 23) recently noted that 'non-governmental organizations... have not only provided information on otherwise neglected scandals, but also sometimes indulge in high-profile public actions – forcing the administrative and enforcement agencies to *do something*'.

As citizen environmental enforcement philosophy changes, INGOs continue to develop and some organisations work within the system by partnering with environmental enforcement agencies, while others work outside of the system to pressure corporations and government to modify their behaviour.

Problems of justice

Environmental justice is a central issue in green criminology (White, 2008). Environmental justice is often defined as:

> [The] fair treatment of all races, cultures, incomes and education levels with respect to the development, implementation, and enforcement of environmental laws, regulations and policies. Fair treatment implies that no population of people should be forced to shoulder a disproportionate share of the negative environmental impacts of pollution or environmental hazards or be denied a proportionate share of the positive benefits of environmental regulation or program environmental hazards due to lack of political or economic strength. (Rhodes, 2005: 8)

We examine the issue of environmental justice to determine if INGO operations and activism are directed from high-income countries, potentially to the detriment of low-income countries. Such a condition suggests environmental injustice exists with respect to the benefits of environmental regulation on the part of civil society INGOs because of a lack of political or economic strength. For instance, INGOs located in high-income countries that are directing volunteers in low-income field offices may create problems because they don't understand local culture. This lack of understanding may shape how effective an INGO will be on the ground (Grossman and Rangan, 2000). It is for this reason that environmental enforcement INGOs that are directed from a high-income country may have trouble addressing ecological disorganisation in a low-income country. For example, McPeak (2001: 478) suggests that the development of INGOs has begun 'to decentralize their management structures... [and is]... accompanied by increasing levels of conflict between field and headquarters staff, falling morale, and proposals to recentralize operational management.'

The distribution of environmental enforcement INGOs has implications for environmental justice through the impact on ecological disorganisation.

The distribution of INGOs may also be related to environmental justice, because INGOs, if largely concentrated in high-income countries, are thought to damage low-income countries:

> In the last years ... a number of Northern NGOs are opening offices in the South, engaging local organisations as representatives or establishing other forms of physical presence. The assumption behind it is that this would make their organisations more efficient and effective, being nearer the target group and the partners in the South. However, there are also some negative views on the move, which has been referred to as a form of neo-colonialism. (Gómez, 2008: 6)

Thus, the location of INGO headquarters in high-income countries may be detracting from civil society in low-income countries. On a global scale, some countries without INGO headquarters may be less likely to form organisations that can engage in environmental justice advocacy as a result of the formation of INGOs in developed countries, because the local organisations may have to compete with organisations in wealthy countries for funding and resources (Stretesky et al., 2011). Thus, the agglomeration of INGOs in high-income countries may shape the global distribution of community environmental policing and advocacy. Low-income countries would then be in a position where policing levels were based on funding directed from foundations and governments in high-income countries. Heavy reliance on foundation funding has been noted as a particularly vulnerable position for nonprofit organisations in general, since it can corrupt grass-roots practices that challenge state authority (Benford, 2005). Again, this condition could detract from environmental justice if the result were that low-income countries were less likely to build the civil society necessary to combat environmental problems.

If the headquarters of environmental enforcement INGOs are largely located in high-income countries, then civil society will develop unevenly across the globe. The uneven development of civil society in the environmental enforcement arena suggests a potential lack of uniformity in environmental enforcement that allows treadmill policies to continue through a 'race to the bottom' mentality. The operations and advocacy aspect of environmental enforcement INGOs should be of specific interest to criminologists, as it has implications for the distribution of criminal justice. Countries where INGOs are headquartered may have more environmental protection and may be determining where pockets of global environmental enforcement are directed. Ideas about power, harm, and justice, then, are directly relevant to green criminology because of the location of headquarters and the implications of that location for civil society.

Data and methods

The purpose of this research is to examine the distribution of environmental enforcement INGO headquarters. Specifically, we studied the distribution of formalised INGOs that engage in environmental enforcement–related activities that include (1) lobbying for the adoption of stricter laws; (2) urging governments to pursue civil and criminal actions for violations; (3) urging governments to pursue stricter penalties or sanctions for violations; and/or (4) promoting better compliance. We collect data on the number, characteristics, founding date, location, and mission of these organisations to determine if environmental enforcement INGOs are more likely to operate in high-income countries than in low-income countries. Such a finding suggests that global environmental protection is dictated by organisations in high-income countries. This finding also raises questions about the effectiveness of combating ecological disorganisation in low-income countries, since organisations in high-income countries are left to identify and prioritise ecological disorganisation. The unequal distribution of environmental enforcement INGOs has implications for environmental justice and the development of a global civil society. We begin our analysis by describing how environmental enforcement INGOs are identified, their function, and how they are distributed over space and time.

Environmental enforcement INGOs

Environmental enforcement INGOs were identified though an examination of all environmental organisations in the *Encyclopedia of International Organizations* (Gale Research Inc., 1997, 2001, 2008), the *Yearbook of International Organizations* (Union of International Associations, 2001, 2005, 2008), and the *World Directory of Environmental Organizations* (International Union for Conservation of Nature and

Natural Resources, 1992, 1996, 2001). To locate environmental enforcement INGOs in those directories, the two researchers examined each environmental organisation within each directory and coded the mission statement or organisation description. Coding was conducted so that if either researcher identified an organisation as an environmental enforcement INGO, it was included in the final list of environmental enforcement INGOs. Although organisational deaths or discontinuations are not examined in this research, most environmental enforcement INGOs had current websites that indicate they still operate.

In this study, we classify environmental enforcement INGOs as those international nonprofit organisations that engaged in (1) traditional enforcement operations and (2) advocacy and lobbying actions.

We defined operations INGOs as

> Organisations that carry out international, national, state, or local legal actions (civil or criminal) to enforce environmental laws, rules, regulations, or agreements and/or obtain penalties or criminal sanctions for violations. This includes monitoring efforts. INGO may also provide direct aid (in the form of resources or monitoring for environmental violations) to state or other governmental agencies that carry out environmental protection efforts.

A total of 126 organisations were identified by the two researchers as meeting this definition of an environmental enforcement INGO. We define lobbying and advocacy organisations as those that advocate or lobby governments (international, national, state, or local) for the creation of enforcement of environmental laws, rules, regulations, or agreements.'

The researchers identified 190 organisations as advocacy and lobbying organisations. Overall, inter-coder agreement reached 95 per cent. This high level of agreement suggests consistency in the coding of INGOs for the purposes of determining whether they engage in environmental enforcement. In the few instances when disagreements in coding did occur, the organisation under consideration was included in the data set, simply in order to capture all potential environmental enforcement INGOs.

In total, 263 organisations are identified as engaging in some type of environmental enforcement (operations, advocacy, or lobbying). Fifty-four of these organisations were classified as engaging in both operations and lobbying and advocacy activities. Table 9.1 presents the number of organisations by country, as indicated in the directories.

Countries not listed in Table 9.1 were identified as having no environmental enforcement INGO headquarters. Thus, 48 countries around the globe contain at least one environmental enforcement INGO headquarters that directs field offices in one or more additional countries. As noted, most headquarters are located in the United States (n = 61) and the United Kingdom (n = 39). According to information obtained from the directory and organisation websites, the INGOs in our analysis were established at various points in time over the course of the last century. Specifically, Figure 9.1 suggests that while a few organisations

Table 9.1 Number of environmental enforcement INGO headquarters, by country*

Argentina	3	Jordan	7
Australia	1	Liechtenstein	2
Austria	4	Malaysia	2
Bangladesh	2	Malta	2
Barbados	2	Mexico	2
Belgium	18	Nepal	1
Bulgaria	1	Netherlands	15
Cameroon	2	Nigeria	1
Canada	14	Norway	2
Colombia	1	Philippines	1
Costa Rica	3	Poland	1
Denmark	1	Russian Federation	4
Ecuador	2	Slovak Republic	1
Egypt	2	South Africa	1
Estonia	1	Spain	2
Fiji	1	Sweden	1
Finland	3	Switzerland	12
France	7	Taiwan	1
Germany	14	Thailand	2
Greece	3	Ukraine	1
Hungary	1	United Kingdom	39
India	3	United States	61
Italy	4	Uruguay	2
Japan	4	Zimbabwe	3
		Total	263

Note: *Environmental Enforcement INGO headquarters are not present in countries that are not listed in Table 9.1.

Source: Data derived from the *Encyclopedia of International Organizations* (Gale Research Inc., 1997, 2001, 2008), the *Yearbook of International Organizations* (Union of International Associations, 2001, 2005, 2008), and the *World Directory of Environmental Organizations* (International Union for Conservation of Nature and Natural Resources, 1992, 1996, 2001).

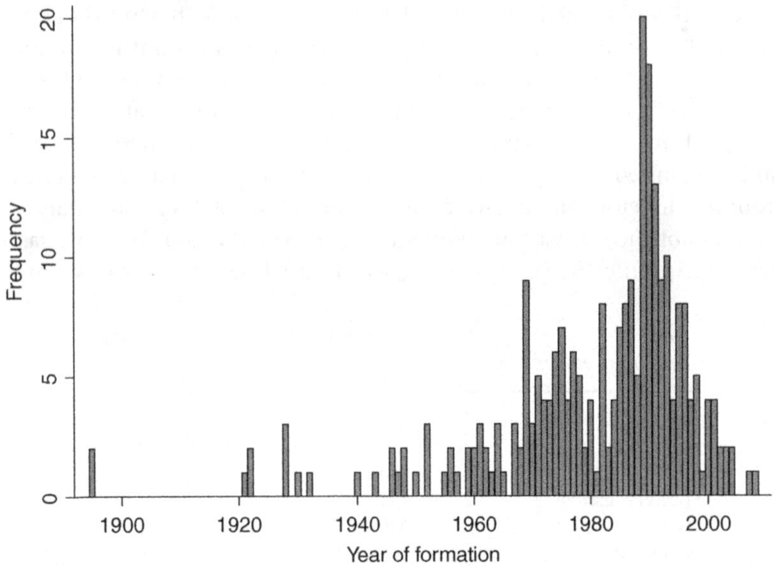

Figure 9.1 Number of INGOs established by year, 1895–2008

Source: Data derived from the *Encyclopedia of International Organizations* (Gale Research Inc., 1997, 2001, 2008), the *Yearbook of International Organizations* (Union of International Associations, 2001, 2005, 2008), and the *World Directory of Environmental Organizations* (International Union for Conservation of Nature and Natural Resources, 1992, 1996, 2001).

were established as early as 1895, most were formed between 1980 and 1999 (n = 148). This observation is consistent with observations about international civil society (Boli and Thomas, 1999).

Income

Are environmental enforcement INGOs headquartered in more wealthy countries than poor countries? To answer this question, we gathered data on per capita gross national income (GNI) for each country for the year 2010. Per capita GNI was the total the value of goods and services produced by a country's economy in 2010, divided by the total population in 2010. Per capita GNI is the main indicator that the World Bank uses to classify world economies. We chose the year 2010 because it represents the current state of the global income distribution relative to the location of environmental enforcement INGO headquarters. The worldwide per capita GNI in 2010 was US $9071. However, the variation in per capita GNI across countries ranges from a low of $170 per person in Burundi to a high of $87,350 in Norway. Data on per capita GNI were obtained from the World Bank (http://www.worldbank.org/) and are measured in thousands of dollars.

Analysis and results

We first examined the average per capita GNI by comparing countries that have one or more environmental enforcement INGO headquarters to countries that do not have any INGO headquarters. The mean per capita GNI in those countries without any headquarters was $5210, while the mean per capita GNI for those countries with at least one headquarters was $13,705. This bivariate comparison suggests that there is an association between country income and the location of environmental enforcement INGO headquarters.

To analyse the association between the number of existing environmental enforcement INGOs and per capita GNI across countries in more detail – and account for the total number of INGOs per country – we use the negative binomial regression (NBREG) command in Stata (Version 11), a commonly used statistical software package. Negative binomial regression is an appropriate statistical technique for modelling count variables (Cameron and Triveldi, 1998). In the case of INGOs, our dependent variable simply represents a count of the number of organisations per country. Because we were also interested in determining the importance of income relative to other variables that might be related to the location of organisations, we controlled for country population size (in hundreds of millions of people) and the total number of international nongovernmental organisations operating within each country (in hundreds of organisations). Research suggests that population size and the existence of other noncompetitive civil society organisations are strongly and positively related to the location of nonprofit organisations (Stretesky et al., 2011). Data on country population and total INGOs were obtained from the World Bank (http://www.worldbank.org/).

Table 9.2 presents the results of the analysis of income and INGOs. Results in Table 9.2 are expressed as Incident Rate Ratios (IRRs) that can be interpreted as the factor increase (ratios above 1.0) or decrease (ratios below 1.0) in the incidence of INGOs across countries that are associated with a one-unit change in the independent variable (that is, per capita GNI, population, or existing NGOs). Thus, in Model 1, Table 9.2, a $1000 increase in per capita GNI is associated with an 8 per cent increase in the expected number of INGO headquarters across countries ($p < .05$). This association is substantive, given the large range in per capita GNI ($170–$87,350).

Table 9.2 Environmental INGOs regressed against per capita Gross National Income, population, and existing NGOs, 2010

Variable	Model 1	Model 2	Model 3
	IRR (SE)[a]	IRR (SE)	IRR (SE)
	....	....	....
Per capita GNI	1.08 (.01)*	1.08 (.01)*	0.99 (.02)
Population	–	1.38 (.26)	1.14 (.14)
Existing No. of NGOs	–	–	1.18 (.04)*
N	171	150	141
Log Likelihood	−181.7	−169.1	−147.9
Pseudo R^2	0.12	0.14	0.20

Notes: $^*p < .05$; [a] Coefficients are expressed as incidence rate ratios (IRR) with standard errors (SE) in parentheses.

Sources: Data derived from the World Bank and the *Encyclopedia of International Organizations* (Gale Research Inc., 1997, 2001, 2008), the *Yearbook of International Organizations* (Union of International Associations, 2001, 2005, 2008), and the *World Directory of Environmental Organizations* (International Union for Conservation of Nature and Natural Resources, 1992, 1996, 2001).

The association between income and INGOs remains unchanged when controlling for population (Model 2, Table 9.2; IRR = 1.08; $p < .05$). However, when adjusting for the number of national nonprofit organisations, the association between per capita GDI and INGOs disappears and is no longer substantively or statistically significant (Model 3, Table 9.2; IRR = 0.99; $p > .05$).

This finding suggests that per capita income is important in the location of environmental enforcement INGO headquarters, because it leads to conditions that promote the establishment of civil society organisations in general. In short, the agglomeration of international civil society is being driven by the existence of civil society organisations. International environmental enforcement organisations that engage in global civil society in the form of environmental enforcement are much more likely to be located in countries where a strong national civil society is already established. As a result, Model 3 in Table 9.2 suggests that across countries, the addition of 100 NGOs is associated with an 18 per cent increase in the incidence of environmental enforcement INGO headquarters. Such a finding might be interpreted as indicating that nonprofit organisations in general help develop the infrastructure and human resources needed to develop global environmental enforcement INGOs. Thus, NGOs may represent the mechanism that explains the relationship between income and

INGOs. Moreover, the correlation (Pearson's *r*) between per capita GNI and the number of NGOs is 0.71 ($p < .05$). Thus, the appropriate causal model for the relationship may be diagrammed as follows:

Income –(+)–> National Civil Society –(+)–> Environmental Enforcement INGO

This model has implications for environmental justice and for the treadmill of production in particular. First, it suggests that the uneven development and agglomeration of civil society within countries has implications for the pattern of development of environmental enforcement organisations across countries. Thus, wealthy countries may have developed the nonprofit infrastructure for global civil society and are driving the decisions about how and where environmental enforcement will be carried out around the globe. The concentration of INGO headquarters in wealthy countries is problematic to the extent that it reflects concerns about neocolonial practices. If solutions to environmental problems are largely isolated in wealthy countries, then the effectiveness of those organisations may be questioned, because organisations in wealthy countries may tend to be more formalised and thus are less likely to question state practices.

In a study of environmental justice organisations in the United States, Rios (2000) found that many environmental justice organisations could be described as formal interest groups as opposed to grass-roots organisations. Larger and more formal organisations often rely on and compete for limited government and foundation monies to operate. This situation may lead to conditions where organisations are more responsive to state and corporate funders than to civil society. Thus, INGOs operating in wealthy countries may not frame environmental protection in the same way as poor countries do, because funding dictates and directs priorities (Benford, 2005). If environmental enforcement INGOs are going to slow ecological disorganisation and the treadmill of production, then those organisations must not be directed by corporate funding and/or state agencies. While there is no guarantee that INGOs headquartered in low-income countries would be free from pressure and corruption, the fact that environmental enforcement INGOs are headquartered in high-income countries raises serious questions about the development of a global civil society that can alter treadmill practices.

It is also possible that organisations that are directed from wealthy countries are not as familiar with the environmental conditions in poor countries and may not be as adept at building the kind of global civil society that will be needed to protect the environment. Thus, the agglomeration of INGOs in wealthy countries may also be less effective at slowing the treadmill of production than if those organisations were more evenly distributed across the globe.

Conclusions

We examined environmental enforcement INGOs and argue that citizen-led environmental organisations that have developed outside the formal state system may be in direct conflict with the state and may put considerable pressure on policy makers and enforcement agencies. As green criminologists, we believe that the distribution of these organisations matters. In the case of environmental enforcement organisations, environmental injustice may occur because environmental 'solutions are not distributed evenly across or within populations' (Gould et al., 2008: 22). We raise social justice concerns that we see as tied to the location of INGO headquarters, and to the location of environmental enforcement organisations in particular. We caution that the potential unintended consequences of the development of INGOs as a response to the treadmill of production may mean that an inequality develops between countries that are relatively poor and those that are relatively wealthy, with INGOs located in wealthy countries dictating where and how enforcement will occur across the globe. Although we find that income is related to the location of environmental enforcement INGOs, it is only indirectly related through existing levels of national civil society. This finding suggests that wealthy countries with a large number of national NGOs are directing environmental enforcement. This is concerning, because studies of more formalised NGOs suggest that nonprofit organisations may be increasingly influenced by state and corporate funders and therefore may have a difficult time challenging state practices.

While international organisations have taken up a major role in the development and enforcement of environmental laws, researchers in criminology have yet to address the distribution of these organisations. We argue that this omission is surprising, because the development of civil society has significant implications for the prevention of ecological disorganisation and crime. The field of green criminology emphasises power, harm, and justice and is well positioned to address these issues. It is our hope that this chapter has encouraged more research on the role of environmental enforcement INGOs, including the role and impact of these organisations in preventing environmental crime and ecological disorganisation.

Notes

1. A few criminologists have talked about the importance of NGOs. For example, the work of Angus Nurse (2011) emphasises the importance of NGOs in wildlife crime.
2. 'Ecological disorganization' is a term used by Schnaiberg (1980). It refers to the disorganisation of matter in the form of natural resources as it is transformed into products used in society. Ecological disorganisation is based on the law of thermodynamics that suggests that matter cannot be produced or destroyed. However, Schnaiberg notes that matter is generally less harmful to health when organised by nature. Production disrupts this ecological organisation because it transforms natural resources into products. The process of transformation of matter not only disrupts ecosystems that rely on the natural order, but also creates pollution that is often harmful to humans and hard to control. Thus, the removal of natural resources and the release of pollution create ecological disorganisation.
3. Environmental justice is concerned with the creation of environmental hazards in the production process as well as the unequal distribution of environmental benefits and/or burdens across diverse races, ethnicities, and classes. The distribution of environmental enforcement is an important environmental justice issue (Stretesky and Hogan, 1998).
4. Toxic chemicals are those chemicals that can lead to injury or death. Various levels of toxicity are generally referenced in relation to the median lethal dose (Lynch and Stretesky, 2011).

References

Agnew, R. (2011) 'The ordinary acts that contribute to ecocide: a criminological analysis', paper presented at the *American Society of Criminology Conference*, 16 November 2011, Washington, DC.

American Lung Association (2010) *State of the Air Report, 2010*, http://www.lung.org/assets/documents/publications/state-of-the-air/state-of-the-air-report-2010.pdf, date accessed 3 July 2012.

Benford, R. (2005) 'The half-life of the environmental justice frame: innovation, diffusion, and stagnation', in D. N. Pellow and R. Brulle (eds) *Power, Justice, and the Environment: A Critical Appraisal of the Environmental Justice Movement*. Cambridge, MA: MIT Press, pp. 37–53.

Boli, J. and Thomas, G. M. (1997) 'World culture in the World polity: a century of international non-governmental organization', *American Sociological Review*, 62(2): 171–190.

Boli, J. and Thomas, G. M. (1999) *Organizations Since 1875*. Stanford, CA: Stanford University Press.

Bondaroff, T. (2011) 'Sailing with the Sea Shepherds', *Journal of Military and Strategic Studies*, 13(3): 1–55.

Brisman, A. (2008) 'Crime-environment relationship and environmental justice', *Seattle Journal for Social Justice*, 6(2): 727–817.

British Petroleum, Inc. (2011) *BP Statistical Review of World Energy*, http://www.bp.com/statisticalreview, date accessed 5 June 2012.

Burns, R. G., Lynch, M. J. and Stretesky, P. B. (2008) *Environmental Law, Crime, and Justice*. El Paso, TX: LFB Scholarly.

Cameron, C. and Triveldi, P. (1998) *Regression Analysis of Count Data*. Cambridge: Cambridge University Press.

Dorn, N., Van Daele, S. and Vander Beken, T. (2007) 'Reducing vulnerabilities to crime of the European waste management industry: the research base and the prospects for policy', *European Journal of Crime, Criminal Law and Criminal Justice*, 15: 23–36.

Eman, K., Meško, G. and Fields, C. B. (2009) 'Crimes against the environment: green criminology and research challenges in Slovenia', *Journal of Criminal Justice and Security*, 11(4): 574–592.

Frank, D. J. (1997) 'Science, nature, and the globalization of the environment, 1870–1990', *Social Forces*, 76(2): 409–435.

Gale Research Inc. (1997) *Encyclopedia of Associations: International Organizations* (31st edn). Detroit, MI: Gale Research.

Gale Research Inc. (2001) *Encyclopedia of Associations: International Organizations* (36th edn). Detroit, MI: Gale Research.

Gale Research Inc. (2008) *Encyclopedia of Associations: International Organizations* (45th edn). Detroit, MI: Gale Research.

Gibbs, C., Gore, M. L., McGarrell, E. and Rivers, L. (2010) 'Introducing conservation criminology: towards interdisciplinary scholarship on environmental crimes and risks', *British Journal of Criminology*, 50(1): 124–44.

Gómez G. M. (2008) *Field Presence of Dutch NGOs: What Is the Impact on Civil Societies in the South?* Institute of Social Studies, http://www.pso.nl/files/2008%20%20field%20presence%20dutch%20ngos.pdf, date accessed 3 July 2012.

Gottchalk, M. (1998) 'Monkeywrenching as punishment', *Criminal Justice Policy Review*, 10: 193–211.

Gould, K. A., Pellow, D. N. and Schnaiberg, A. (2008) *The Treadmill of Production: Injustice and Unsustainability in the Global Economy*. Boulder, CO: Paradigm Publishers.

Greenaway, F. (1992) 'Foreword to the reprinted edition', in A. Clow and N. Clow (eds) *The Chemical Revolution*. Philadelphia, PA: Gordon and Breach Science Publishers, pp. xi–xii.

Grossman, A. and Rangan, V. K. (2000) 'Managing multi-site nonprofits', *Harvard Business School Social Enterprise Series, Number 8*, http://www.hbs.edu/socialenterprise/pdf/SE8ManagingMultiSiteNonprofits.pdf. Accessed 4 July 2012.

International Union for Conservation of Nature and Natural Resources (1992) *World Directory of Environmental Organizations* (4th edn). Claremont, CA: California Institute of Public Affairs.

International Union for Conservation of Nature and Natural Resources (1996) *World Directory of Environmental Organizations* (5th edn). Claremont, CA: California Institute of Public Affairs.

International Union for Conservation of Nature and Natural Resources (2001) *World Directory of Environmental Organizations* (6th edn). Claremont, CA: California Institute of Public Affairs.

Kobayashi, J. (2009) 'Making the connections: water, forests, and minerals exploitation in South and Southeast Asia', in R. Cronin and A. Pandya (eds) *Exploiting Natural Resources: Growth, Instability, and Conflict in the Middle East and Asia*. Washington, DC: Henry L. Stimson Center.

Lynch, M. J. (1990) 'The greening of criminology: a perspective for the 1990s', *The Critical Criminologist*, 2(3): 3–4, 11–12. Reprinted in P. Beirne and N. South (eds) (2006) *Green Criminology*. Aldershot, UK: Ashgate, pp. 165–170.

Lynch, M. J. and Stretesky, P. B. (2003) 'The meaning of green: contrasting criminological perspectives', *Theoretical Criminology*, 7(2): 217–38.

Lynch, M. J. and Stretesky, P. B. (2011) 'Similarities between green criminology and green science: toward a typology of green criminology', *International Journal of Comparative and Applied Criminal Justice*, 35(4): 293–306.

Matus, K., Kyung-Min, N., Selin, N. E., Lamsal, L. N., Reilly, J. M. and Sergey P. (2012) 'Health damages from air pollution in China', *Global Environmental Change*, 22(1): 55–66.

McPeak, M. (2001) 'Tackling fragmentation and building unity in an international nongovernmental organization', *Nonprofit Management and Leadership*, 11(4): 477–491.

Nurse, A. (2011) 'Policing wildlife: perspectives on criminality in wildlife crime', *Papers from the British Criminology Conference*, 11, 38–53. Available at http://www.britsoccrim.org/volume11/pbcc_2011_wholevolume.pdf#page=42. Accessed 6 July 2012.

Obach, B. K. (2004) 'New labor: slowing the treadmill of production?' *Organization and Environment*, 17(3): 337–54.

Oliver, W. (1998) *Community Oriented Policing: A Systematic Approach to Policing*. Upper Saddle River, NJ: Prentice-Hall.

O'Rourke, D. and Macey, G. (2003) 'Community environmental policing: assessing new strategies for public participation in environmental regulation', *Journal of Policy Analysis and Management*, 22: 383–414.

Overdevest, C. and Mayer, B. (2008) 'Harnessing the power of information through community monitoring: insights from social science', *Texas Law Review*, 86: 1494–1526.

Palmer, M. A., Bernhardt, E. S., Schlesinger, W. H., Eshleman, K. N., Foufoula-Georgiou, E., Hendryx, M.S., Lemly, A. D., Likens, G. E., Loucks, O. L., Power, M. E., White, P.S. and Wilcock, P. R. (2010) 'Mountaintop mining consequences', *Science*, 327(5962): 148–149.

Pellow, D. (2004) 'The politics of illegal dumping: an environmental justice framework', *Qualitative Sociology*, 27(4): 511–525.

Rhodes, E. L. (2005) *Environmental Justice in America: A New Paradigm*. Bloomington: Indiana University Press.

Rios, J. M. (2000) 'Environmental justice groups: grass-roots movement or NGO networks? Some policy implications', *Policy Studies Review*, 17: 179–211.

Samet, J. M., Zeger, S. L., Dominici, F. C., Coursac, I., Dockery, D. W., Schwartz, J. and Zanobetti, A. (2000) *The National Morbidity, Mortality, and Air Pollution Study. Part II: Morbidity and Mortality from Air Pollution in the United States*, Research Report 94. Cambridge, MA: Health Effects Institute.

Schnaiberg, A. (1980) *The Environment: From Surplus to Scarcity*. New York: Oxford University Press.

Schorn, D. (2005) 'Burning Rage', *60 Minutes*, http://www.cbsnews.com/2100–18560_162–1036067.html. Accessed 4 July 2012.

Situ, Y. (1997) 'A pathway to the knowledge of environmental crime: learning through service', *Journal of Criminal Justice Education*, 8(2): 243–351.

Skogan, W. G. and Harnett, S. M. (1997) *Community Policing, Chicago Style*. Oxford: Oxford University Press.

Smith, J. and Wiest, D. (2005) 'The uneven geography of global civil society: national and global influences on transnational association', *Social Forces*, 84(2): 621–652.

South, N. (1998) 'A green field for criminology? A proposal for a perspective', *Theoretical Criminology*, 2(2): 211–233.

Srivastava, L., Schwartz, S. L. and Austin, M. J. (2012) *Managing Non-Governmental Organizations Worldwide: Mapping the Knowledge Base of Nonprofit Management in the Human Services*, unpublished manuscript, http://www.mackcenter.org/docs/NGOs%20Worldwide.pdf. Accessed 4 July 2012.

Stretesky, P. B. (2008) 'The neglect of race and class in environmental crime research', in M. J. Lynch and R. Paternoster (eds) *Racial and Ethnic Bias in the Criminal Justice System*. New York: Criminal Justice Press.

Stretesky, P. B., Huss, S., Lynch, M. J., Zahran, S. and Childs, B. (2011) 'The founding of environmental justice organizations across US counties during the 1990s and 2000s: civil rights and environmental cross-movement effects', *Social Problems*, 58(3): 330–360.

Stretesky, P. B. and Lynch, M. J. (2011) 'Coal strip mining, mountaintop removal, and the distribution of environmental violations across the United States, 2002–2008', *Landscape Research*, 36(2): 209–230.

Stretesky, P. B., Shelley, T. O. and Crow, M. S. (2010) 'Do conservation organizations influence the social production of natural resource violations?' *Organization and Environment*, 23(4): 398–416.

Union of International Associations (2001) *Yearbook of International Organizations* (35th edn). Geneva: Union of International Associations.

Union of International Associations (2005) *Yearbook of International Organizations* (42nd edn). Geneva: Union of International Associations.

Union of International Associations (2008) *Yearbook of International Organizations* (45th edn). Geneva: Union of International Associations.

Vanderheiden, S. (2005) 'Eco-terrorism of justified resistance? Radical environmentalism and the war on terror', *Politics and Society*, 33: 425–447.

White, R. (2008) *Crimes against Nature: Environmental Criminology and Ecological Justice*. Abingdon, UK: Willan.

Woods, N. D. (2006) 'Interstate competition and environmental regulation: a test of the race-to-the-bottom thesis', *Social Science Quarterly*, 87(1): 174–189.

World Health Organization (2011) *Media Centre Fact Sheet: Air Quality and Health*, http://www.who.int/mediacentre/factsheets/fs313/en/index.html. Accessed 4 July 2012.

Yale Law School (2011) *Environmental Law*, Career Development Office, http://www.law.yale.edu/Body_2011_PUBLIC.pdf, date accessed 4 July 2012.

10
A Harm Analysis of Environmental Crime

Diane Solomon Westerhuis

Introduction

One of the greatest challenges for the study of environmental crime is the analysis of harms against the environment, and how these are prosecuted (or not) and sanctioned. Bricknell (2010: 117) proposed the need for a comprehensive analysis of a specific environmental harm and 'the array of current and potential preventative, enforcement and punishment responses'. White (2008) describes the need for a scoping analysis of environmental harms, and then reminds us that 'adequate data collection and analysis forms part of what is needed if environmental prosecution and sentencing is to move forward' (White, 2010: 379). White notes the exception of New South Wales, because there is now a sentencing database available from the New South Wales Land and Environment Court (NSWLEC). This chapter takes up that challenge and describes a study of environmental harms and responses in that Australian specialist court. The NSWLEC has considered environmental planning and protection to be criminal enforcement matters since the inception of the court in 1980.

In this study I undertook a survey of the range of crimes prosecuted recently in the NSWLEC, in order to analyse the types of environmental crime that are prosecuted in this jurisdiction. The crimes are studied to assess environmental harms and how they are described and sanctioned in the court. The diversity of sentencing and the sanctions imposed, particularly attempts at reparation or amelioration of the harms inflicted, are examined, and an attempt is made to understand just what sort of justice is achieved. The aim of the study is to attain some understanding of environmental harms and of our responses–in particular, the power this specialised court has to achieve what may be considered environmental or green justice. The focus of this chapter is on environmental harms, and although it may seem to be a recent concern, the history of legislative response is not new. We can begin with narratives about earlier environmental harms and legislative responses.

The most persistent harm we inflict was and is that of pollution: the harm we inflict upon the biosphere and the environment in which we operate. We humans have been persistent and efficient at pollution from a very early period. Probably the earliest record we have of a response to such pollution is a proclamation of King Edward I of England (1272–1307), who had became concerned over pollution from coal burning. Edward used the law to attempt to change the behaviour of the community; he banned the burning of 'sea-coal' in fireplaces and authorised the destruction of furnaces and kilns (Brimblecombe, 1976: 946). The sentence for offenders allegedly was torture or execution; whether anyone was actually put to death cannot be confirmed. There have been many proclamations and legislation regarding pollution since then, yet still the burning of coal continues (see Freese, 2006; Whitehead, 2009), and we find new ways to pollute.

Legislation to criminalise such activities is not a new response, and is often used in conjunction with or as an alternative to regulation. Although various monarchs and states since Edward's time have passed diverse laws, and regulations have continuously increased in attempts to reduce harms, air pollution and its associated harms have persisted. In London, pollution became particularly detrimental during the Industrial Revolution, which eventuated in the deaths of many people in big industrial cities. London 'pea-soupers' became notorious; in 1952 a four-day 'Black Fog' killed approximately 4,000 Londoners (Freese, 2006: 168). The response to this was the enactment by Parliament of the Clean Air Act 1956 (United Kingdom, repealed 1993), when legislation was used to attempt to persuade people to change their behaviour. The penalty for offenders is stated in the Act, section 27(3): 'a fine not exceeding one hundred pounds or to imprisonment for a term not exceeding three months or to both'. In the United Kingdom, and in most other developed countries, pollution is now an environmental crime and the harms it produces are now penalised, yet clearly air pollution and other pollutions continue.

In attempts to control pollution, most states regulate and legislate, circumscribing where waste and polluting materials can be dumped, which then criminalises the dumping of waste in other locations. Yet the locations which are designated as dumping grounds for waste, or where other dangerous activities such as factories, power stations, or nuclear reactors are allowed, are not always safe for humans or other animals and may also impinge upon neighbouring land or cross borders. While such activities may be legal, the location of these licensed activities could often be described as unjust or discriminatory and could produce calls for environmental justice. Environmental justice is a concept that originated in critiques of the location of hazardous material or the distribution of environmental goods such as access to clean water in locations that disadvantage vulnerable groups. Recent examples of environmental injustice include the location of hazardous waste or pollution in habitats occupied by minority groups or first peoples, or the destruction of their environments for industry such as logging. Examples occur in remote or disadvantaged communities, as Bullard (1990) describes in 'Dumping in Dixie', where polluting industries were located in poor communities in the Southern states of the United States. Particularly disturbing examples are identified in a United Nations report in 2012 – for example, that of the indigenous Akuntsu people of Brazil, whose population was wiped out in the 1980s by illegal logging activities. At the time of writing (2012), five of the tribe have survived, they now live in a tiny patch of forest surrounded by ranches (Nellemann and Interpol Environmental Crime Programme, 2012: 24). Other examples in the report identify where illegal logging is endangering the lives of indigenous peoples of the tropical forests of the Amazon, Central Africa, and Southeast Asia, extreme cases of environmental injustice.

The concept of environmental justice is clearly one that has diverse applications and entails notions of environmental harms and the power to control or reduce harms inflicted by environmental crimes. The concept has broadened to encompass access to justice in the courts and theories of ecological justice, which consider 'non-human parts of the environment as deserving protection for the environment's sake and not just because of human interest and utility' (Mann, 2010: 217). It is this aspect of environmental or ecological justice that is of interest here. Justice in these cases must include responses to harms and, just as in other deliberations of justice, consideration of the victim, particularly when the victim is 'the environment', or some aspect of the environment such as the rivers or seas, the air, the biota, or non-human animals. The responses to harms to these victims are instrumental in achieving any sort of environmental justice, and so are examined in the cases that appear in the NSWLEC.

Legislation and the NSWLEC

With few exceptions, environmental crime in New South Wales (NSW) has not been considered as extreme or of great interest until relatively recently. Norberry argues that until the 1980s, 'penalties have historically been low and there has been little in the way of enforcement of pollution laws' (Norberry, 1995: 2). I would place the watershed a little earlier, when licences to pollute were issued under diverse NSW statutes such as the Clean Waters Act 1970, the Clean Air Act 1961, and the Pollution Control Act 1970. In 1979, these were bought together under the new Land and Environment Court Act of 1979, the enabling legislation of the NSW Land and Environment Court (NSWLEC), in which a series of licensing regulations were bought under the surveillance of the NSWLEC, under section 17.

This was followed during the 1980s and since then with more complex environmental legislation in response to community concerns, interwoven through federal, state, and local government legislation and regulations. Prior to the Protection of the Environment Operations (POEO) Act 1997 (NSW), sentencing in court cases in the NSWLEC was based on the Environmental Offences and Penalties Act (EOPA) 1989 (NSW), which set out penalties for environmental crimes for individuals and corporations in three tiers, from the most serious, which allow penalties not exceeding \$1 million for corporations and \$250,000 for individuals, and/or a maximum 2 years or (if taken to the Supreme Court) 7 years imprisonment. The POEO Act continued this regime, but increased penalties substantially: Tier 1 offences can result in penalties of up to \$5 million and 7 years in gaol, although custodial sentences are extremely rare. Tier 2 offences of failing to notify a pollution incident include a maximum penalty of \$2 million in the case of a corporation and \$500,000 in the case of an individual. The maximum penalties for Tier 2 offences other than failure to notify about pollution incidents are \$1 million in the case of a corporation and \$250,000 in the case of an individual. Further daily penalties apply to continuing offences. Lesser offences are dealt with by penalty notices or 'on-the-spot fines'. However the court now under section 250 of the POEO Act also has wider power to order expenses and compensation to a public authority or a victim, to pay costs occurred by prosecution or by other regulatory authorities, the payment of any profit made by the offender, and to 'name and shame' – that is, the publication of the offense at the offender's cost. Other options include various remedies: specified projects for the restoration or enhancement of the environment in a public place, to pay monies to the Environmental Trust or other environmental organisation, and education and training courses for offenders, employees, or contractors. Of particular interest is that the Act includes tradable emissions schemes and green offsets.

Cases discussed here also refer to other Acts, including the Environmental Planning and Assessment Act 1979 (NSW), the Environment Protection and Biodiversity Conservation Act 1999 (Cth), the Native Vegetation Act 2003 (NSW), the National Parks and Wildlife Act 1974 (NSW), and the Threatened Species Conservation Act 1995 (NSW). I propose that the increasing complexity of such regulations and legislation regarding environmental crime is evidence of two things. The first is the increasing concern in the community about environmental harms – for example, with the growth of the Greens parties as a political voice, which policy makers and legislators must increasingly take into account. The second far outweighs the first; it is the ideology of economic rationalism, which produces policies and legislation that frame any prosecution of environmental harms in the context of licences, permits, and markets. Tradable emissions and green offsets are evidence of this market-based philosophy. The outcome is that now we have regulation and legislation that licenses some to pollute and to harm the environment (regulated) and others to harm the environment without a licence, of whom only some are prosecuted.

The data

The majority of cases in the NSWLEC are civil, regulatory, noncriminal cases, where arbitration is used to settle disputes for example, over development, removal of trees, or mining leases. Cases are classified as civil or criminal, for arbitration (classes 1–4), appeals and mining disputes (Class 6), or criminal cases (Class 5). Recently, mining cases were added in Classes 7 and 8 appeals and are also included in Class 5 cases. As we have seen in the Acts discussed above, the court allows a great range of legal enforcement, which includes negotiated noncriminal outcomes, traditional sentencing outcomes, and alternatives, including self-reporting and self-regulating. It also includes fines, noncustodial and (very rarely) custodial sentences, plus many alternative sanctions, which attempt to control or remediate harms. Justice Stein describes this as a 'shift with increasing environmental awareness for more social control' (Stein, 1995: 4).

The creation in NSW of an environmental crime sentencing database using data from the NSWLEC was a response to the concerns amongst the judiciary for consistency and transparency in sentencing (Preston and Donnelly, 2008). The database includes the range of penalties available and sentences imposed in each case, as well as the various considerations taken into account when sentencing, including the degree of environmental harm, the history of the offender, and the offender's culpability, as well as admission of guilt and other variables. Although subscription is costly for researchers, the database is a useful tool for legal research, particularly for common law. At a much more accessible price is the free, government-funded Australasian Legal Information Institute (AUSTLII) database, which allows free access to the full transcript of each reported case, from which the data for our present research was accessed. Reported cases are usually uploaded and transcripts made available on the AUSTLII site within a couple of weeks of the decision hearing. Other sources of information used in this research include NSWLEC Annual Reports and speeches by justices, available on the NSWLEC website.

Overall, the number of criminal Class 5 cases prosecuted in the NSWLEC has grown steadily since the inception of the court in 1980, with a substantial peak around 1990 (see Walters and Solomon Westerhuis, 2013). With the enactment of the Environmental Offences and Penalties Act 1989 (NSW), prosecutions in the Land and Environment Court 'rose from 40 in 1988 to 193 in 1989 and 317 in 1990' (Stein, 1995; Pain, 1995: 7). This peak has never been reached again. Class 5 cases disposed of by hearing increased gradually from 68 in 2006 and in 2007, to 94 cases in 2009, but dropped to 47 in 2010 (NSWLEC, 2011: 26). This study focuses on the most recent cases.

The convenience sample for this research consists of 100 cases located in a search of all sentencing decisions in reported cases available in AUSTLII, heard in the NSWLEC, described as Class 5. Cases where there was no case to answer, where the charges were dismissed, where defendants were found not guilty and therefore no sentence was imposed, hearings for costs, contempt cases or appeals, cases with duplicate case numbers, and cases where final decisions have not yet been reached were excluded. The search was extended back from the time of the research, September 2012, until 100 cases fitting the criteria were located, the earliest in June 2007. The selected transcripts were searched for all evidence of the crime committed, the harm caused, how it was described, and the sanctions imposed. Particular attention was paid to any remediation attempted or rehabilitation required in order to reduce or remedy the harm inflicted. Table 10.1 shows the number of cases identified that fit the criteria.

Table 10.1 Sample of NSWLEC Class 5 cases from AUSTLII

Year	Cases, n = 100
2012	14
2011	18
2010	23
2009	28
2008	9
2007	8

Offences

The range of offence types (see Table 10.2) is of interest to an understanding of what environmental harms were perpetrated. By far the most frequent offence in this sample is the pollution of waters (30 per cent), which included creeks, rivers, and oil pollution in the sea. Corporate offenders appear most frequently, particularly in the pollution of waters, which, with the exception of a council, were all companies or corporate offenders. Below I discuss some of these offences and the sanctions ordered by the NSWLEC; see Table 10.2.

Harms such as pollution, as we have seen, have long been accepted as criminal, and case law is well established. However, they are very diverse crimes and pollution describes a great many different types of harm. In this sample, we can include air pollution, water pollution, the dumping of waste and sewage where it should not be, many of them described as a 'breach of environment protection licence condition'.

In the case of *Environment Protection Authority v Delta Electricity* [2009] NSWLEC 11 (11 February 2009), the defendant holds an Environment Protection Licence for a power station, which includes a repository for fly ash, the byproduct of burning coal for electricity production. However, the corporation had failed to minimise or prevent the emission of fly-ash dust, which is a condition of its licence. Thus it had breached one of the environment protection licence conditions, the failure to minimise or prevent the emission of dust from its premises. While under the POEO Act the maximum penalty of $1 million is possible, there was in this case a substantial reduction in penalty in light of numerous mitigating circumstances, and the court ordered a fine of $45,000 plus costs of $35,000.

Matters were more serious in the case of *Environment Protection Authority v Causmag Ore Company Proprietary Limited* [2009] NSWLEC 164

Table 10.2 Offences prosecuted in AUSTLII sample of 100 cases

Offences	Per cent
False accreditation – carrying out a statutory site audit when not accredited	1
Breach of environment protection licence – other	1
Clearing native vegetation	13
Damage to habitat of threatened species	3
Damaging reserved land	1
Harming threatened species animals	1
Picking an endangered ecological community/picking endangered population of plants/picking threatened plant species	8
Polluting air	8
Polluting waters/marine/oil	30
Pollution with sewage	2
Pollution with waste	8
Threatened species damaged by council's road works	1
Tree clearing/pruning in breach of tree preservation order/without consent	8
Without consent/development consent breach/excavation works without consent	13
Damage Aboriginal object property of Crown	1
Demolition of local heritage	1
Total	100

(24 September 2009), in which the offender, a mining company, had two previous convictions for similar offences. In this case, the company breached its environment protection licence condition by failure to maintain filter bags in a proper and efficient manner in order to prevent escape of dust, and the court considered this a reasonably serious failure. The resulting harm was considered reasonably foreseeable, even though there appeared to be no environmental harm other than short-term amenity impacts. Clearly this case was considered by the court in light of the previous offences, that the defendant was recidivist, and so the court ordered not only a fine of $20,000, plus costs of $26,500, but also a publication order and an order for contributions to an environmental project. This is an example of the more innovative and alternative sanctions now being applied in the NSWLEC. The company was ordered to place an advertisement in the first 12 pages of the local paper. In addition the defendant was ordered to pay an environmental fine to the local council for a landfill project; the fine consisted of $15,000 each year in monthly instalments for the next three years, totalling $45,000 on top of the court fine and costs. Any references made by the company to the

donations to the landfill project were to be accompanied by a notice describing the offence and a declaration that their remedial actions were ordered by the court.

Water pollution cases are more frequent than air pollution in this court. In *Environment Protection Authority v Forgacs Engineering Pty Limited* [2009] NSWLEC 64 (30 April 2009), a shipbuilding company was guilty of a breach of environment protection licence condition; in this case, it failed to prevent he emission of particles from its floating dry dock. It was considered by the court (at 28) that although there 'is no evidence of actual ecological harm ... there was potential for some harm from the tributyltin (TBT) floating in the water and finally settling on the bottom of the harbour, meaning a real, not remote, possibility, but given the lack of specific data in the field it is impossible to quantify' and that 'harm can be cumulative so that activities which contribute incrementally to the gradual deterioration of the environment should be treated seriously'. Steps were taken to mitigate and prevent further harm, the causes of which were apparently foreseeable. The court ordered a publication notice and a contribution to an environmental project. The amount of $45,000 was to be paid to the local council for the purpose of replacing 250 m of timber decking over mangroves in a boardwalk upgrade, to provide environmental education access to mangroves, and to provide access for maintenance and litter removal. In addition, any references by the defendant to its funding of the project was to be accompanied by a notice explaining that the funding was a condition of its sentence when it had been convicted of breaching a condition of its environmental protection licence, and how this had occurred.

In the case of *Environment Protection Authority v Moolarben Coal Operations Pty Ltd* [2012] NSWLEC 65 (30 March 2012), a mining company was convicted after multiple discharges of pollutant-sediment-laden waters into a creek, contra to their conditional approval for coal mining. The harm was described as reasonably foreseeable, and the defendant was fined the sum of $105,000, plus prosecution and investigation costs of $61,632. In addition, there was a publication order, intended to both shame the offender and educate the general public about the offence and penalty.

Environment Protection Authority v Queanbeyan City Council (No 3) [2012] NSWLEC 220 (18 September 2012) is of particular interest, as the city council was convicted of polluting the local river 'and waters downstream thereof' across a border from NSW into the adjoining Australian Capital Territory (ACT). The operations of the Protection of the Environment Operations Act 1997 (NSW) allowed the court to consider such environmental harms extraterritorially, that is, where harms imposed in the next state cross the state border. While there had been no intent and limited environmental harm, there would have been an increased risk of exposure to viruses and protozoa. There was prior criminality and a late plea of guilty. The court imposed a substantial costs order of $343,000, a publication order, plus an environmental services order with its usual declaration of guilt. In this case the council was ordered to pay the local Catchment Management Authority the sum of $80,000 to be used for a landscape project.

Pollution cases appear well supported by expert evidence and relatively identifiable harms, and community expectations are clear that these crimes will be pursued to prevent adverse health outcomes and protect community safety. In contrast, decisions about land clearance and chopping down of trees were in the past actions commonly undertaken by the landholder, and they did not need to be permitted or licensed by the state. These actions are now regulated in Australia, and in NSW clearing vegetation of all types is monitored and subject to regulation. Native vegetation is considered significant for ecologically sustainable development and so clearing is described as a key threatening process affecting the survival of threatened species. Clearing in rural areas is subject to the Native Vegetation Act 2003 (NSW), which requires either development consent or a property vegetation plan; in urban areas, the Environmental Planning and Assessment Act 1979 (NSW) requires development consents and is subject to tree preservation orders. However this has not been an easy transition; controversy has surrounded some cases and success is questionable. Indeed Bartel (2003: 116) complains about the lack of enforcement:

> The aims of the regulations are unlikely to be achieved. Too much land is approved for clearance and post-clearance revegetation works are favoured at the expense of protecting remnants. Monitoring is heavily tree-centred and implementation is suffering due to a pragmatic but nonetheless self-defeating political response to stakeholder influence. Satellite data shows that land clearance has declined in New South Wales. If this is due to the prosecution of the early cases then there is little reason for it to decrease any further once it becomes known that the biggest implement in the enforcement toolbox is no longer being used.

Since Bartel was writing in 2003 of these concerns, land clearing has been in the news and a substantial number of cases have appeared in the courts. In this sample, cases of clearing native vegetation included one case each of clearing koala habitat, clearing squirrel glider habitat, clearing reserve land (native vegetation) and picking plants (golden wattle), picking endangered ecological communities or endangered population of plants or threatened plant species (8 per cent), and tree clearing without consent (8 per cent).

Contrary to the perceptions about land clearing in the literature, it is not necessarily landholders or farmers who are the offenders. In some cases, the focus is on the licence and on the breaching of conditions by corporate defendants, and the focus is not always on the harm caused to the environment, but rather the adherence to regulation or licensing. For example, in *Minister for Planning v Moolarben Coal Mines Pty Ltd* [2010] NSWLEC 147, Justice Craig noted: 'It is the harm to the integrity of the planning system which lies at the heart of the offence in this case' (at 73) as the offender had failed to seek approval before carrying out work. While the claim was made that it was the integrity of the planning system that was harmed, the clearing of 3.5 hectares of native vegetation, of which at least 1.3 hectares was described as endangered ecological communities (EEC), was identified as 'objective harmfulness' by the court. In this case, the defendant suggested to the court that the harm was minimal, yet this could not be assessed, because no prior environmental assessment had been carried out, as should have been undertaken if planning requirements had been met. The court concluded that (at 59) 'there was harm in the short to medium term from the loss of vegetation, particularly the loss of the EEC, with the ecological values incidental to it, albeit that the harm was ultimately assessed by the prosecutor as being minimal'.

A significant factor in sentencing was deterrence. Justice Craig in *Minister for Planning v Moolarben Coal Mines Pty Ltd* [2010] NSWLEC 147 (at 46) denounced the

> corporation engaged in a multimillion dollar coalmining project, an activity which in all its aspects had the potential to have a very significant impact upon the environment. Corporations engaged in activities of this kind must be reminded of the obligations imposed upon them to ascertain the laws and controls applicable to the carrying out of any activity associated with a particular project in order to ascertain the need for any consent or approval to that activity so that the appropriate environmental assessment can be undertaken before any approval is given to it. The sentence therefore needs to make clear to the community at large that the failure

> of a corporation to take these steps will be visited with significant financial penalties.

The offender submitted that after the offense it had undertaken a planning application modification and had volunteered substantial offsets, works, and undertakings that were, according to the opinion of its consultant, in excess of the requirements to compensate for the loss of vegetation and particularly for the EEC. The defendant submitted that it would be appropriate to impose an order for tree planting and maintenance rather than a fine, a suggestion rejected by Justice Craig, who pointed out that this commitment to offsets was in compliance with the conditions already imposed for licensing, and therefore the defendant was already obliged to undertake the tree planting and maintenance. The fine imposed was $100,000, to which a discount of 30 per cent was applied for the utilitarian plea of guilty and assistance to the court, resulting in a fine of $70,000, plus costs of $55,000.

Another significant case focussed on endangered species, in *Plath v Chaffey* [2009] NSWLEC 196. The offender, an amateur bird enthusiast and oological (bird egg) collector, visited Lord Howe Island, but was apprehended leaving the island with eggs of four threatened species and two protected species of birds. It was noted by the court that the primary consideration in sentencing is the objective gravity or seriousness of each offence. In determining the objective gravity or seriousness of each of the offences in this case, the circumstances that the court considered included the fact that Mr Chaffey had not applied for a licence, and so he had

> offended against the legislative objectives expressed in the statutory offences and thwarted the achievements of the objects of the Act, including ecologically sustainable developments. In respect of four of the charges, the animals were of vulnerable species and also were components of populations of those species of conservation significance. The animals of both the threatened species and protected fauna were components of a world heritage area. Mr Chaffey's conduct has caused actual environmental harm (at 29).

Commenting on the nature of Mr Chaffey's offences, Justice Preston noted that the defendant would be unable to pay a fine; ability to pay and financial circumstances are often considerations when sentencing orders are made. Harm to the environment was considered by the court, as well as the need in sentencing for reparation to the community for the environmental harm caused by Mr Chaffey's conduct in collecting the eggs. Justice Preston ordered a community service order of 80 hours, the hours calculated by the number of eggs of each of the species harmed. The options for community service were to be undertaken at the local showground or museum, not in the habitat harmed.

Harms and victims

Although in other courts there may be different outcomes, in the NSWLEC the majority of criminal cases heard were about pollution (48 per cent), followed by vegetation clearance (21 per cent); see Table 10.3. In Table 10.3, the integrity of the planning system is listed as a victim in 16 per cent of the cases, but many cases listed under other harms are also concerned about the integrity of the system, including, for example 'contravening a condition of a threatened species licence'. This priority differs somewhat from what appears to be of interest in the literature. Recent research published on environmental crime in Australia, particularly that published by the Australian Institute of Criminology, focuses in the main on environmental crime involving fishing and timber industries (White, 2007, 2008a; Putt and Anderson, 2007; Schloenhardt, 2008). The exception is the overview by Bricknell (2010), which attempted to synthesise all environmental crimes and includes a brief summary of not only the timber and fishing industries but also of pollution, water theft, illegal trade in fauna and flora, and clearing of native vegetation. In my sample there was one case of water theft, which was not included in the final analysis as the sentence had not yet been ordered. There were no cases of illegal trade, although there were many cases of harm to habitats and the one case of the theft of eggs, as discussed above.

A broad comparison of the crimes prosecuted in the NSWLEC (see Table 10.2) allows us to consider what could be described as the victims of the offences in our sample (Table 10.3). The data were examined to establish what had been harmed, or what could be described as a victim, in each of the 100 cases, and the results were collated in Table 10.3. This is very much a generalisation adduced from the descriptions of harm described in the transcripts, but the categorisation allows us to consider just what is being harmed, according to the court.

We would expect in environmental crime cases that the environment would be the victim most often, but it can be seen from the data in Table 10.3 that the integrity of the planning system is also the victim, just as much as threatened species. However the perception remains in the court that pollution, tree clearing, or clearing native

Table 10.3 Categories of victims of offences

Victim	Offences	Per cent	Total per cent
The integrity of the planning system	Breach of environment protection licence – other	1	
	Without consent/development consent breach/excavation works without consent	13	16
	Damage to Aboriginal object, property of Crown, demolition of local heritage	2	
Threatened species	Contravening a condition of a threatened species licence	1	
	Damage to habitat of threatened species	3	
	Damaging reserved land	1	
	Harming threatened species animals	1	
	Picking an endangered ecological community/picking endangered population of plants/picking threatened plant species	8	15
	Threatened species damaged by council's road works	1	
Native vegetation	Clearing native vegetation	13	
	Tree clearing in breach of tree preservation order/without consent	8	21
Pollution	Polluting waters/marine/oil	30	
	Polluting air	8	
	Pollution by sewage	2	48
	Pollution by waste	8	

vegetation – particularly without consent – and any sort of contravening of licences are very much the victims of harms. Pollution is clearly most often the harm inflicted upon the environment, particularly when we consider other offences such as 'use of land as waste facility without lawful authority/transporting waste', which because it is without licence, also harms the integrity of the system. Many of the pollution crimes were about where people dumped waste, without a licence or contra to a permit. This is clearly a major concern of the courts, particular in the breach of licence, or breach of permits or conditions of approval. Justice Biscoe stated in *Minister for Planning v Coalpac Pty Limited* [2008] NSWLEC 271 (11 September 2008) (at 44): 'At stake is the integrity of the planning system which is harmed when a person carries out development before undertaking the required assessments and receiving approval'. The offence in this case was that a mining company had 'carried out development under a project approval contrary to condition 6 of schedule 2 of the approval in that it produced more than 350,000 tonnes of saleable coal in a year', for which they were fined $200,000 plus costs.

Sanctions

The NSWLEC has the power under the Protection of the Environment Operations (POEO) Act 1997 (NSW) to order diverse sanctions. The diversity of sentencing (see Table 10.4) is of interest for a discussion regarding harms and how sanctions may make attempts at reparation or remediation. Fines are the most frequent sanction, but they do come in two different varieties: the fine paid to the court or, instead what is described as an environmental fine, paid to a particular agency or for a particular purpose, often an environmental project. Examples of agencies include National Parks and Wildlife, Catchment Authorities, or local councils for particular projects.

The largest court fine ordered was in the case of *Director-General of the Department of Environment and Climate Change v Hudson* [2009] NSWLEC 4, a fine of $400,000 for the offence of clearing native vegetation without a development consent. It was noted by the court (at 62) that while the land owners 'hold the land in fee simple and they regarded the trees as theirs, they nevertheless remain subject to

Table 10.4 Sanctions for Class 5 sentencing cases*

Sanction	Total	Per cent of Cases
Fine	$4,787,100	78
Investigation costs	$234,377	26
Environmental fine	$1,587,600	23
Environmental projects	Varied projects, in addition to those ordered by environmental fine	13
Publications	Most often with environmental fine or project	31
Mea culpa notice	Often accompanies environmental fine or project	14
Community service hours	1580 hours	4

Note: *100 percent of cases were ordered to pay prosecution costs, in addition to any sanction imposed.

laws passed by the State parliament and which apply to all citizens within the State, including the Native Vegetation Act, which secures the sustainable management and conservation of native vegetation.'

Justice Lloyd considered this case within the upper range of seriousness, but found that the landowner was 'somewhat misguided as to what he is able to do on his land' (at 91) and so reduced the maximum penalty of $1.1 million to a penalty of $400,000, plus costs, for one of the two charges. There was no further penalty, other than costs, and no order in this case for any environmental project or fine. Prosecutors' costs were ordered in all cases in our sample, but in 26 per cent of cases, investigation costs were also ordered.

Environmental projects include specific directions by the court for an environmental audit or a remediation plan, which included the costs of hiring consultants and planning remediation in specific areas that had been harmed, or the planting of trees to replace those destroyed. There were in this sample few attempts to clean up pollution, and there were 10 cases where orders were made to plant trees, monitor habitats, remedy harm or, in one specific case, to create a more friendly environment. In *Environment Protection Authority v Tea Garden Farms Pty Ltd* [2012] NSWLEC 89, the defendant was found guilty of polluting water in a marine park and was ordered to fund two environmental restoration and enhancement projects. The first was to a Bushland Reserve Project for stabilisation and remediation of tracks within the reserve; the second was to the Marine Parks Authority in the Marine Park for the installation of 'seagrass-friendly moorings' in place of 'dump and chain' swing moorings, as described to the court, clearly designed for 'alleviating harm to the marine environment'. Cleanup costs ordered could be quite substantial; for example, in *Environment Protection Authority v Buchanan* (No 2) [2009] NSWLEC 31, there were prosecutor's cleanup costs of $88,395.75, plus the other cleanup costs.

The recognition of harm done to the victim of the crime and the community was attempted in many cases, but not always. For example, in *Chief Executive, Office of Environment and Heritage v Coffs Harbour Hardwoods Sales Pty Ltd* [2012] NSWLEC 52, the defendant had cleared 21 Newry golden wattles (*Acacia chrysotricha*), an endangered species listed in the NSW Threatened Species Conservation Act. The court found that the wattle trees were damaged or killed, along with a number of other protected native trees and plants in the reserve, leading to approximately 4,000 square metres of the Nature Reserve being cleared to bare earth. Justice Lloyd noted that (at 39) 'In this case the victim and the community are the threatened species involved and the community interest in preserving the threatened species and endangered species, and the purposes of having a nature reserve'. In this case the defendant was ordered to undertake a five-year plan for weed control and extensive remediation in the National Park.

A specific attempt at alleviating harms was evident in the case of *Chief Executive of the Office of Environment and Heritage v Bombala Investments Pty Ltd* [2012] NSWLEC 115, in which both Bombala Investments Pty Ltd and the director of the company were charged with harm to the habitat of the squirrel glider (*Petaurus norfolcensis*), listed as vulnerable. Bombala Investments Pty Ltd was fined $13,000, and the director fined $10,000, to be paid to the National Parks and Wildlife Fund specifically for mapping and study of the squirrel glider populations in the habitat harmed, and in addition the defendants were required to retain a bush regenerator, an ecologist and an expert with special knowledge of the threatened squirrel glider species, to prepare a remediation plan for the affected areas and to carry out the plan for regeneration of cleared vegetation and any other actions recommended by the consultants. Publication was required of the offence by both defendant and company, and costs to be paid by both.

While this attempt at remediation is specific to the site and habitat of the victim, other sanctions included payment of monies to more general environmental projects (see Table 10.4) in attempts to remedy harms, and in 4 per cent of cases, offenders were ordered to undertake community service, in another form of reparation to the community. Such sanctions appear common to restorative justice and may be more common in the future (see Preston, 2011). Clearly in most cases deterrence is the focus; accountability and denunciation are also evident.

Orders for publication can have a direct effect on the offender, and sometimes this is perceived as a more direct penalty than a fine. In the case of *Environment Protection Authority v Coastal Recycled Cooking Oils Pty Limited* [2008] NSWLEC 242 (at 43), the defendant made a plea to the court regarding the impact that publication would have on his business: 'The impact on the Defendant's business in terms of customer perception might be severe given that its business is an environmentally worthwhile activity of recycling used cooking oil for biofuels. The cost of the advertisement alone of up to $1,000 is substantial.' The court did not order publication in this case.

The undertaking of environmental projects, combined with publications in local newspapers or trade publications, such as the *Australian Mining Monthly Magazine*, plus declarations of guilt, are designed to shame the offender and educate the public about the offence, and perhaps work towards deterrence. In other cases, the court ordered that all future public references by the defendant to the project were to contain a disclaimer, or what could be called a 'mea culpa', describing the offence and the penalty. For example in *Environment Protection Authority v George Weston Foods Ltd* [2010] NSWLEC 120 (at 89.6), the orders stated that

> All future references by the defendant to its funding of the Peel River Riparian Project Stage 2 must, pursuant to s 250(1)(a) of the *Protection of the Environment Operations Act* 1997, be accompanied by the following passage:
>
> 'George Weston Foods Limited's funding of the Peel River Riparian Project Stage 2 is part of a penalty imposed on George Weston Foods Limited by the Land and Environment Court after it was convicted of polluting waters, namely the Peel River, at Tamworth, NSW, an offence against s120(1) of the *Protection of the Environment Operations Act* 1997.'

This declaration of guilt accords well with restorative justice principles, as the notice apportions blame and therefore shame.

However in 46 percent of cases a fine and costs were the only sanction, and nothing was attempted to acknowledge or remediate any harms. More often, concern was expressed about the system, of great concern to the court. One case stands out in this category, that of *Environment Protection Authority v Djura* [2012] NSWLEC 122 (29 May 2012). In this case, the defendant had presented himself as accredited site auditor, and carried out a bogus statutory site audit of potentially contaminated land, and pleaded guilty to offences under sections 57(1) and 48(1)(a) of the *Contaminated Land Management Act* 1997 (NSW) (the CLM Act). Under the CLM Act, matters to be taken into account when considering a penalty for offences include the harm caused or likely to be caused by the actions. Although the judge noted that this is not a case where the offences have caused environmental harm, the judge found that there is 'a need to protect the integrity of the accredited site auditor scheme. The scheme is in place to ensure that risks to human health and the environment from contaminated land are properly identified and [that] appropriate steps taken to remediate that land. If persons who are not appropriately qualified purport to carry out site audits, then the integrity of that system is undermined and there is a risk of harm through inappropriate use of land. The defendant's conduct undermined the integrity of the system' (*Environment Protection Authority* v *Djura* [2012] (at 50)).

The harm then was to the system, and the defendant was fined $7,500 plus costs of $12,000. The judge decided that, for the purposes of sentence, particularly (a) punishment, (b) deterrence, (e) accountability, and (f) denunciation were relevant in this case.

Discussion

Watson (2005: 199) argues that 'environmental crime will remain profitable until the financial costs to offenders outweigh the likely gains. The anticipated net benefit of environmental crime to offenders must become negative.' Perhaps the fines in this sample do achieve their objective of deterrence. In surveying the cases in this sample, it can be noted that few defendants are repeat offenders. Deterrence may well be an outcome of the sentencing regime of the NSWLEC. Although we have not seen any great reduction in the number of criminal offences since the high of the 1990s, which would be one indicator, we do not see many recidivists; those who have been penalised in the past rarely offend again. In this sample, those who do re-offend are in the main corporate offenders; they should therefore, on a second offence, lose any licence they hold to pollute. However the maximum fine has not been ordered in any of these cases.

Overall from this study we have identified the most frequent types of offences and have found that sanctions are most frequently about deterrence. Remediation is clearly attempted in the cases described above; however, any attempt at remediation or at addressing environmental harms was attempted in only 54 per cent of cases, and in many of these cases, this consisted of only a publication or of an environmental fine on top of the court fine and costs.

Recommendations for policy considered as a consequence of this research appear straightforward. If we are to achieve any sort of environmental justice, we must reduce the number of offences that harm vegetation, species, air and water, as White suggests (2008a), and we must remedy harms that we cannot prevent. As we have seen in this sample, sanctions applied in the NSWLEC go some way towards this, particularly those that do include an environmental project as remedy; however, the court has power to do much more. In 46 per cent of cases, an opportunity for remedy was not taken up. Sanctions could contribute to at least one of three remedies; the first – the withholding of permits or licences, including those permitting companies or individuals to pollute – was rarely contemplated. Cleaning up pollution was only ordered specifically as cleanup costs in one case; cleaning up pollution or planting native vegetation could become common remedies undertaken by both individuals and corporate offenders, even if these are not harms caused by the original crime. Although this may seem simplistic, it would go a long way towards restoration of the harms caused and would achieve some semblance of environmental justice.

References

Bartel, R. L. (2003) 'Compliance and complicity: an assessment of the success of land clearance legislation in New South Wales', *Environmental and Planning Law Journal*, 20(2): 116–141.

Bricknell, S. (2010) *Environmental Crime in Australia. AIC Reports Research and Public Policy Series 109*. Canberra: AIC.

Brimblecombe, P. (1976) 'Attitudes and responses towards air pollution in medieval England', *Journal of the Air Pollution Control Association*, 26(10): 941–945.

Freese, B. (2006) *Coal, A Human History*. London: Arrow Books.

Land and Environment Court of NSW (2011) *Annual Review 2010*. Sydney: NSW LEC.

Mann, T. (2010) *Australian Law Dictionary*. Melbourne: Oxford University Press.

Nellemann, C. and INTERPOL Environmental Crime Programme (2012) *Green Carbon, Black Trade: Illegal Logging, Tax Fraud and Laundering in the Worlds Tropical Forests. A Rapid Response Assessment*. United Nations Environment Programme. Available at www.grida.no.

Norberry, J. (1995) 'Environmental crime', in N. Gunningham, J. Norberry and S. McKillop (eds) *Environmental Crime: Proceedings of a conference held 1–3 September 1993, Hobart*. Canberra: Australian Institute of Criminology.

Pain, N. (1995) 'Criminal law and environment protection – overview of issues and themes', in N. Gunningham, J. Norberry and S. McKillop (eds) *Environmental Crime: Proceedings of a Conference held 1–3 September 1993, Hobart*. Canberra: Australian Institute of Criminology.

Preston, B. J. (2011) 'The use of restorative justice for environmental crime', *Criminal Law Journal*, 35(3): 136–154.

Preston, J. B. J. and Donnelly, H. (2008) *Achieving consistency and transparency in sentencing for environmental offences, Research Monograph 32*. Sydney: Judicial Commission of New South Wales.

Putt, J. and Anderson, K. (2007) *A national study of crime in the Australian fishing industry. Research and public policy series no. 76*. Canberra: Australian Institute of Criminology.

Schloenhardt, A. (2008) *The illegal trade in timber and timber products in the Asia-Pacific region. Research and public policy series no. 89*. Canberra: Australian Institute of Criminology.

Stein, P. (1995) 'The role of the NSW land and environment court in environmental crime', N. Gunningham, J. Norberry and S. McKillop (eds) *Environmental*

Crime: Proceedings of a Conference held 1–3 September 1993, Hobart. Canberra: Australian Institute of Criminology.

Walters, R. and Westerhuis, D. (2013) 'Green crime and the role of environmental courts', *Crime, Law and Social Change*, 59(3): 279–90.

Watson, M. (2005) 'Organised crime and the environment: the British experience', *European Environmental Law Review*, 14(8/9): 207–213.

White, R. (2008) *Environmental harm and crime prevention. Trends and issues in crime and criminal justice no. 360*. Canberra: Australian Institute of Criminology.

White, R. (2010) 'Prosecution and sentencing in relation to environmental crime: recent socio-legal developments', *Crime, Law and Social Change*, 53(4): 365–381 DOI: 10.1007/s10611-010-9233-x.

Whitehead, M. (2009) 'The wood for the trees: ordinary environmental injustice and the everyday right to urban nature', *International Journal of Urban and Regional Research*, 33(3): 662–681.

11

Victims of Environmental Harms and Their Role in National and International Justice

Matthew Hall

Introduction

Over the last two decades, increasing attention has been paid by criminologists to the natural environment and to criminal activities that lead to environmental degradation (White, 2008). Researchers in the field that has been variably labelled 'conservation criminology' (Gibbs et al., 2010), 'eco-critical criminology' (Lynch and Stretesky, 2007), and, more recently, 'green criminology' (Ruggiero and South, 2010) have made steady progress towards the application of criminological theories of offending and crime prevention to such activities. From a critical perspective, criminology has also cast light on the power imbalances inherent in the labelling of certain polluting activities as 'criminal', which of course is tied up with the economic goals of corporate actors and indeed of states as a whole (Pepper, 1993). Throughout this development, however, green criminology has repeated the omission only now being fully recognised within mainstream criminology: excluding *victims* of crime from such academic discourse. Over the last 40 years, the sub-discipline of victimology has gathered pace and to some extent has addressed this shortfall in relation to more 'traditional' notions of victimisation (property crime, crimes of violence, domestic violence, and so on). However, there is at present an almost complete absence of victimological work focusing on those affected by environmental crime, or indeed focusing on the wider concept of environmental harm.

In this chapter I will set out a research agenda for the better understanding of so-called 'environmental victimisation'. The varying impacts of environmental harm on individuals and communities will be addressed and the present limits of national and international justice systems in addressing these needs will be discussed. I present the case for an interdisciplinary approach to these issues, that encapsulates criminology, victimology, and international law.

'Environmental harm' and 'environmental victims'

The failure of mainstream victimology to account for those falling victim to environmental harm appears to derive from a number of interrelated features of this particular kind of victimisation. The first is that, even when one is considering officially recognised environmental offences set down in legislation, environmental victimisation poses a number of novel problems for the traditional criminal justice system in most jurisdictions. One is that such offences may often involve a large group or community of victims, perhaps with competing interests (Skinnider, 2011), whereas most criminal justice systems are designed around single offenders and single victims. Furthermore, perpetrators of such victimisation may be corporations, or even states, which again poses challenges to traditional models of criminal justice, which are designed to tackle individual offenders (Giddens, 1990). In addition, it is often difficult to draw the necessary lines of causation (to the criminal standard of proof) between perpetrator and victim, leading some to dismiss environmental crime as 'victimless' (White, 2011).

Clearly then, environmental victimisation does not fit neatly within standard conceptions of victimisation now being employed by most criminal justice systems in the wake of the wider 'victims movement' (see Hall, 2010). Perhaps more fundamental, however, is the fact that many of the activities that foster environmental victimisation are not officially proscribed as 'crimes' at all in the majority of jurisdictions. As noted by Skinnider (2011: 2):

> Many environmental disruptions are actually legal and take place with the consent of society. Classifying what is an environmental crime involves a complex balancing of communities' interest in jobs and income with ecosystem maintenance, biodiversity and sustainability.

Passas (2005) makes the point more succinctly when he labels the majority of polluting activities as 'lawful but awful'. Critical brands of victimology have long since recognised that what we define as 'criminal' and therefore what we define as 'criminal victimisation' is heavily influenced by power imbalances in society (Hough, 1986). Nevertheless, for McBarnet (1983), it is the victimologists themselves who are partly to blame for this state of affairs. By concentrating their attention predominantly on traditional notions of victimhood, McBarnet suggests that researchers in the field have played into the hands of governments wishing to derive political capital from victims and from punitive criminal justice responses:

> Victimology has contributed to the strengthening of the state's role. It has set itself up as engaging not just in academic debate but in 'affirmative action for the victims of crime', and, like traditional criminology before it, its too ready acceptance of official definitions of criminal and victim have reinforced rather than questioned the status quo. (McBarnet, 1983: 302)

I submit that this argument gains particular weight when applied to environmental harm, because such harms may be facilitated (or at least endorsed) by the state itself, and in some cases might even amount to state crime (Green and Ward, 2004). Commentators such as Elias (1983, 1986) and Rock (1990) have gone further to argue that society's selective and narrow definition of crime, and therefore of criminal victimisation, is politically motivated, and in the case of environmental degradation, we might also add 'economically motivated.' Underlying this restrictive approach is also the fact that victimology as a sub-discipline has come to focus almost exclusively on *criminal* victimisation. In contrast, Pointing and Maguire (1988) describe how the victims' movement in the United States was originally driven by a host of 'strange bedfellows' concerned with 'social' victimisation in a much broader sense (Young, 1997; Garkawe, 2004).

The difficulty with taking such a legalistic approach to environmental victimisation is that it impedes critical discussion of the labelling of certain activities as criminal in the first place and, perhaps more significantly, results in the failure to ascribe the 'criminal' label to many instances of environmental harm. For example, in one of the few existing contributions to this debate specifically focused on environmental victims, Williams (1996: 35) offers the following definition of this group: 'Those of past, present, or future generations who are injured as a consequence of change to the chemical, physical, microbiological, or psychosocial environment, brought about by deliberate or reckless, individual or collective human act or omission.'

In presenting this definition, Williams draws upon the concept of 'injury', as opposed to the wider notion of 'harm', because for him 'governments are more likely to respond in relation to tight, manageable definitions, which may be stretched a little, than to "catch all" concepts that might appear to carry a host of hidden ramifications' (Williams, 1996: 205). This argument is a logical and pragmatic one; nevertheless the purposeful underestimation of environmental victimisation by academics also seems to substantiate McBarnet's (1983) criticism.

For that reason, this chapter takes a broader approach to the issue of environmental victimisation, understanding environmental victims to encompass those harmed by the adverse effects of environmental degradation perpetrated or brought about by individuals, corporations, and states. In adopting this perspective, I am applying what Hillyard and Toombs (2003) call a 'social harms' approach, which for them brings a number of advantages and serves to reinvigorate what they view as a waning critical discourse amongst criminologists. For example, the authors note that 'crime', as argued by Hulsman (1986), has no 'ontological reality' and hence 'the criminal law fails to capture the more damaging and pervasive forms of harm' (Hillyard and Toombs, 2003: 12). As such, focusing on harm has the potential to include the often legally ambiguous activities that foster environmental damage. Indeed, even when such activities are criminal in the strict legal sense, focusing on harm allows us to account for such activities in cases where whatever mechanisms of justice available (at the national, transnational and international levels) fail to adequately prosecute such transgressions. Another salient point made by Hillyard and Toombs is that the social harms approach allows for the consideration of 'mass harms'. Again this chimes well with the problems inherent in man-made environmental degradation, where many thousands of people might be affected. Traditional criminology, on the other hand, has struggled to fully embrace the concept of mass victimisation and, with the exception of limited inroads into the fields of state crime and corporate crime, has largely remained focused on the individual.

In relation to environmental harm, Hillyard and Toombs's approach also has much resonance with some of the earliest literature from what has been termed 'the environmental justice movement' (Williams, 1996: 200). Environmental justice has been variously defined and is generally acknowledged as a wide concept that emphasises the involvement of people and communities in decisions that might impact upon their environment, defined broadly to include their 'cultural norms, values, rules, regulations and behaviours' (Bryant, 1995: 6; see also Hofrichter, 1993 and Čapek, 1993).

Examining environmental victims

Skinnider (2011) discusses a number of ways one can seek to classify victims of environmental harm: by wrongful act; by the nature of the harm; by the extent of the damages suffered; by the scope of the harm, or by the perpetrator(s) of that harm. In keeping with the social harms approach discussed above, the present chapter ostensibly focuses on classifications of harm, as this typology supports the argument that a key difficulty faced by environmental victims is a lack of recognition of the harms they suffer as *criminally* perpetrated. In this section, I discuss four broad classifications of such harm: health impacts; economic impacts; social and cultural impacts, and reduced security. I then go on to expand on two key features of environmental victims – namely, the overlap between victims and offenders and the unequal distribution of such victimisation.

Health impacts of environmental harm

Perhaps the most obvious consequences of environmentally destructive activities for many victims are health implications. Indeed some categories of 'health impacts' are relatively obvious and physically verifiable, albeit perhaps only after an extended length of time. This renders such effects a much better 'fit' with existing legal principles in most criminal jurisdictions around the world, which tend to favour positivistic virtues like certainty, predictability, and objectivity. Such harms are also generally speaking more quantifiable, which aligns them well with systems already in place at the national and international levels to compensate parties physically and mentally injured as a result of crime (Miers, 1997; Hall, 2010). From a legal perspective then it seems we are in fairly recognisable territory when we consider the health implications of environmental degradation, and indeed this category of impact fall within Williams's notion of 'injury' discussed above.

In the criminological literature, Lynch and Stretesky (2001) have analysed the question of corporate harm and violence, utilising evidence from medical literature and related studies that focus on the health consequences associated with exposure to toxic waste, pesticides, and dioxin. In so doing, they argue that the significant health consequences associated with modern industrial production of toxic waste products 'can be thought of as "criminal" in the broadest sense since alternative, nontoxic methods of production are often available' (Lynch and Stretesky, 2001: 153). They also make a point to be taken up in greater detail below: that the health impacts of such dumping are not evenly distributed around the world's population, and in fact tend to fall disproportionately on the already impoverished.

In recent years, the focus of many such inquiries has been on the impacts on human health of distinct man-made environmental 'disaster events' like the Deepwater Horizon oil spill in the Gulf of Mexico in 2010. The human health implications of this event – both physical and mental – are now the subject of a rapidly escalating scientific literature (Lee and Blanchard, 2010; Yun, Lurie and Hyde, 2010). Another prominent example is the negative long-term health impacts of the 1983 Bhopal gas leak in India, including respiratory and neurological disorders, which have been demonstrated by Cullinan, Acquilla, and Dhara (1996). These long-term effects are in addition to the estimated 8,000 people who reputedly died in the immediate aftermath of the disaster itself (D'Silva, 2006). The Chernobyl nuclear disaster of 1986 has of course led to very long-term health implications for those affected at the time, and for those subsequent generations who have lived in the area since. In this case, long-term health impacts identified by the World Health Organisation (2006) include a prevalence of leukaemia, thyroid cancer, and increased mortality.

Of course, whilst disasters such as these have received a great deal of media attention, the crucial point is that these more visible examples of environmental victimisation are inevitably a tiny minority of all those suffering health complains as a result of environmental harms. Patz et al. (2000), for example, report on the long-term and varied consequences of climate change as a whole. Another pertinent example is the legal and illegal dumping of hazardous waste materials, where the health implications of such activities are, if anything, more directly palpable. Ruggiero and South (2010) cite numerous cases of death and illness brought about in areas exposed to hazardous waste materials, including the so-called 'cancer villages' of China, where residents' increased susceptibility to several classifications of tumours has been directly attributed to their exposure to cadmium and mercury released through the recycling of e-waste (Watts, 2010: 21). More recently, the United Nations Environment Programme has reported that in Nigeria, in 'at least 10 Ogoni communities where drinking water is contaminated with high levels of hydrocarbons, public health is seriously threatened' (United Nations Environment Programme, 2011), following 50 years of oil operations.

Even though the negative heath implications are therefore clear in many cases, the difficulty from the perspective of criminal law is to draw the relevant chain of causation between polluting activities and these effects, assuming that the said polluting activity is illegal in the first place. Indeed, such activities may in fact be state sanctioned and even actively encouraged by the state, which has certainly been the case in Nigeria, given the reliance of the country's economy on the oil industry.

Economic impacts of environmental harm

The best estimates of the monetary/fiscal cost of environmentally damaging activities worldwide run into the billions. As such, following on from the increased mortality rates highlighted in the previous section, the WHO recently estimated that deaths caused by air pollution are costing economies within the European Union around €161 billion a year (British Broadcasting Corporation, 2007). The United Nations Environmental Programme (Mullier, 2010) estimates the worldwide turnover of environmental *crime* at $31 billion annually.

Studies have indicated a likely negative impact of climate change on a wide variety of industries, ranging from paper production (Jaggi and Freedman, 2006), the wine industry (Nemani et al., 2001), and tourism (Berrittella et al., 2006), to fishing (Possnert, Tooley and Mörner, 2004). Another example of the broad economic impact of environmental degradation is that of the insurance sector. In 2009, the International Association for the Study of Insurance Economics acknowledged that climate change would inevitably lead to higher costs, 'largely due to socio-economic factors such as value concentrations in coastal areas' (42). In some cases, such negative impacts on industry will have significant financial implications at the national level. For example, in one case study Reid et al. (2007) conclude that climate change will have a major impact on the Gross Domestic Product of Namibia.

In all such cases, a threat to any national or local industry is a threat to the livelihoods and, in many cases, ways of social and economic life of those involved in those industries. As Lee has noted in the context of communities polluted by dioxin around the An-shun plant in southern Taiwan:

> The research findings underscore that poverty can be a serious issue for residents in polluted areas. The reasons for unemployment among able bodied household breadwinners may include pollution-related illness, the loss of markets for local products due to pollution or stigmatization, or the migration of business away from the community. (Lee, 2009: 27)

Even though the economic effects of environmental degradation appear to fall disproportionately on poorer countries (see below), people in more developed parts of the world face similar threats to their means of economic sustenance. For example, in one recent report, a fisherman in Louisiana, facing reduced shrimp hauls ostensibly as a result of the 2010 Gulf Oil spill, was quoted as saying: 'We don't have millions of dollars sitting in the bank where we can go do something else. We live and die on the seafood industry. This is our culture...This is how we live' (Lee, 2011).

It is at this point that the impact becomes not just economic, but social and cultural, again illustrating the holistic nature of the issue. Indeed, increased poverty due to economic downturns precipitated by environmental degradation will almost certainly feed back as negative health implications for those who are impoverished (Murray, 2006). Again, the differing 'impacts' of environmental harm prove difficult to distinguish.

Social and cultural impacts of environmental harm

Quantifying 'social' or 'cultural' damage to a people or community as a result of environmental harms is extremely challenging, although as an exercise it is by no means alien to more mainstream criminology (Dolan and Moore, 2007). As a category of impact, it is also central to the notion of environmental justice mentioned above. These are not abstract speculations; loss of one's traditional cultural activities and lifestyle can itself have significant economic and health effects. For example, there are a number of discussions in the literature concerning the people of the Maldives, who are presently facing significant risk to their homes, economy, and traditional ways of life as a result of sea level rises ostensibly brought about by climate change (Brown et al., 1997; Mörner, Tooley and Possnert, 2004; Possnert et al., 2004; Domroes, 2001). Brown et al. have reflected at length on whether corporate entities or even foreign states might be held responsible,(criminally or otherwise, under international law for the damage that has been done to the islanders' traditional fishery culture, which is of particular relevance to our present discussion (Mörner et al., 2004). The further example of shrimp fishing in the Gulf of Mexico has already been discussed.

The key point for our present purposes is that it is these traditional cultures that also provide these environmental victims with the practical necessities of living (food, livelihood, and so on), as acknowledged by the 2012 Rio+20 United Nations Conference on Sustainable Development, when it stated in its Outcome Document: 'Many people, especially the poor, depend directly on ecosystems for their livelihoods, their economic, social and physical well-being, and their cultural heritage' (Paragraph 30).

The above notwithstanding, the harms brought about by cultural and social damage to a community or an individual extend beyond those that can be followed through to a more tangible impact. In one telling example, Wheatley (1997) has elaborated on the social and cultural impacts of mercury pollution on aboriginal peoples in Canada. Wheatley stresses the holistic view of the environment taken by such communities and notes that the impacts of such harm therefore go well beyond that which can be expressed (or redressed) in monetary terms. Similar observations have been made in the US context, where Brook (1998) has labelled the threat to Native American sovereignty precipitated by the industrial dumping of toxic waste on tribal lands as a form of 'environmental genocide'.

A further dimension to the loss of cultural and social stability brought about by environmental victimisation is the far-reaching criminogenic implications, leading to further victimisations. One especially relevant issue, with which the criminal justice agencies of most developed countries are already heavily concerned, is that of human trafficking. The link between displaced peoples/forced migrations and human trafficking has been expressly drawn by a number of researchers (see Lee, 2007). The United Nations University's Institute for Environment and Human Security (Warner et al., 2008) in particular has demonstrated specific connections between migrations forced by environmental factors and a susceptibility of these displaced individuals to human trafficking. Jasparro and Taylor (2008) have also discussed the links between climate change, culture, and the threat of human trafficking, with a particular focus on South Asian livelihoods.

In the European Union too, the expectation of human trafficking is already a high concern for Member States (Shelley, 2007). I submit that the effects of climate change and other environmental degradation are likely to create a further pull in the direction of stringent collaborative action, with implications for the human rights of those trafficked, many of whom end up working in illegal and poorly regulated sectors of the economy.

Impacts of environmental harm on security

In recent years, ideas regarding 'security' have been increasingly linked to environmental concerns to produce a distinct literature on 'environmental security' (see Hough, 2012). Definitions of environmental security differ (see Heckler, 2011), but generally the concept tends to link environmental degradation and the associated scarcity of resources with human conflict at individual, group, and state levels. Brunnèe (1995) conceives of it as 'the prevention and management of conflicts precipitated by environmental decline' (1742). Although typically limited to the field of armed conflicts (which naturally result in considerable loss of life and personal injury to human victims), more recent definitions of environmental security tend to include a wider body of threats to the natural environment (Ullman, 1983). For example, in recent years the concept of environmental security has led some commentators to speak of 'environmental terrorism', which Chalecki (2002: 3) defines as 'the unlawful use of force against in situ environmental resources so as to deprive populations of their benefit(s) and/or destroy other property'. For the purposes of this present discussion, the important observation is that, as natural resources become restricted by environmental degradation (sometimes legal, sometimes illegal), this is likely to make such resources increasingly precious to states and therefore increasingly attractive to terrorist groups. The response of governments is likely to be increased regulation and the rollout of harsher penalties (and new crimes) for environmental terrorists, just as the scope of 'terrorism' itself was expanded in many jurisdictions in the light of the 2001 terrorist attacks in the United States (Mythen and Walklate, 2006).

Whilst the human impact of threats to environmental security in general are very real, for many they are perhaps less than the dangers posed by the more specific threat to 'food security'. Food security has been defined by the World Food Summit of 1996 as existing 'when all people at all times have access to sufficient, safe, nutritious food to maintain a healthy and active life' (United Nations Food and Agriculture Organization, 2010). The concept is usually understood as including both physical and economic access to food that meets people's dietary needs as well as their food preferences (Pinstrup-Andersen, 2009). At present much of the literature and policy attention in various countries has focused on the immediate health and humanitarian implications of food security coming under threat; however, the legal and criminogenic implications are also beginning to be assessed. MacLeod et al. (2010), for example, have written at length on the introduction of regulative frameworks intended to preserve food security. In China, the National People's Congress Standing Committee has recently introduced criminal sanctions, including heavy fines and prison sentences, to anyone prosecuted for adding poisonous or harmful ingredients to foods (China Network Television, 2011).

Recently Lobell, Schlenker, and Costa-Roberts (2011) have confirmed that the impact of climate change on crop levels is leading to a rise in the price of food, with obvious implications for food security. Of course, a rise in the price of food itself has many criminogenic and victimogenic implications. Lack of food may lead to localised violence and riots about food prices, as demonstrated by the unrest felt across some 20 countries in 2008, when world food prices reached crisis levels (Ivanic and Martin, 2008). In the African context, Takemura (2007) asserts that there is a 'deepening anger and resentment among people at the bottom of society, fostered by a rise in food prices, which could threaten stability in developing countries' (273).

Indeed, the violence that has been predicted to come as a result of food insecurity may well have already found expression, with some suggesting that food prices helped to trigger the unrest in Tunisia and Egypt in early 2011. Historically, food riots are also not alien to the United Kingdom, where they occurred in the eighteenth century (Thompson, 1991), or to the United States, where the 1862 'Bread Riots' were precipitated by droughts, leading to a reduction in grain and other basic foodstuffs, exacerbated by the pressures of the Civil War (Steinberg, 2008).

Food security is also a good example of the often indistinguishable quality of the impacts of climate change and other environmentally harmful activities on the wider environment on the one hand and on human victimisation on the other. Mares, for example, emphasises the knock-on effects of soil erosion brought about by changes in temperature and reduced rainfall:

> [The exhaustion of soil resources across the globe has great potential to create substantial future human insecurity and harm. Most of the negative effects are likely to be found in poorer, developing nations. (Mares, 2010: 283)

Environmental degradation therefore undermines security in a number of ways, which have the potential to exert major consequences for human beings in terms of their health, safety, and continued prosperity. It can also be gleaned from the above that threats to security may prompt increased deviance and criminal activities. It is to this link between victimisation and offending/deviant behaviour that I turn in the next section.

Environmental victims and environmental offenders

It has been well established by victimologists that, far from the 'ideal' stereotype of the blameless, innocent victim (Christie, 1986), real victims of crime in fact often overlap as a group with offenders. Indeed,

Farrall and Maltby (2003) have argued that this reality impedes any divorcement of 'victimology' from the wider ambit of criminology. It is becoming increasingly clear that the same overlap occurs when one considers environmental victims. Take, for example, the potential victims of human trafficking, discussed above. Clearly, as is the case with many trafficked individuals, such victims will inevitably find themselves engaged in illegal activities in the 'receiving' country or region, be that activity prostitution or illegal working. Marmo and La Forgia (2008) comment on how official and unofficial authorities in Australia have a tendency to characterise trafficked women as undesirables, regardless of the circumstances that left them in their present situation and regardless of whether such people are actively engaged in illegal activities. Smith (2007) has noted that some countries have deployed their military forces in reaction to the flow of displaced human populations, perceiving them as a security threat.

Further examples of the overlap between 'environmental victims' and 'environmental offenders' might include food suppliers who are tempted to adulterate their product in times of reduced harvests and rising costs, and therefore may be seen as victims of their economic circumstances (Mandalia, 2005). Indeed, those who protest or even riot as a consequence of a lack of food or high food prices are arguably only doing so as a result the harms that environmental degradation has visited upon them. Clearly, given such a situation there is a concern that the poorest people will turn to illegal food markets. As well as being criminal in themselves, such illegal markets will undoubtedly be run by those who will be willing to use threats or actual violence to ensure that they get the price they want for the goods they sell, and who may also be involved in other allied trades (such as the supply of weapons, drugs, and the control of prostitution), which can only prompt still further victimisations.

> In addition to those turning to activities that have for the most part always been criminalised, increased regulation of the environment, discussed above, can expand the official ambit of 'deviance', in a process criminologists often label 'net widening' (McMahon, 1990), to make new offenders out of the victims of ecological change. Here White (2011) draws on the work of Duffy (2010) to give an example of the creation of wildlife reserves in parts of Africa: 'When wildlife reserves are established, local communities can suddenly find that their everyday subsistence activities have been outlawed and they have been redefined as criminals'. (White, 2011: 113)

The issue of how environmental regulation may expand the power of the state, bringing those to its attention who previously would not have merited official sanction, is a key concern for the critical school. White's point also challenges the assumption that 'the law' is necessarily the only or best way to deal with such 'transgressions'.

Unequal distribution of environmental victimisation

Perhaps the most important observations to be made concerning environmental victims is that the overriding evidence now points to endemic inequality in the distribution of the harms discussed in this chapter (Dobson, 1998). Whilst this inequality is partly dictated by geographical issues (low-lying islands being especially at risk from rising sea levels, for example), it is imperative that this focus on the inequalities of environmental harm fostered by physical geography not distract us from the more complex – social, economic, cultural and political – aspects of environmental victimisation. In fact, we have already noted a number of times in this chapter that the impacts of environmental degradation (both internationally and within countries) is distributed very unevenly, with the poorest, most disadvantaged countries and groups within countries tending to suffer most (International Association for the Study of Insurance Economics, 2009). Lee (2009) has summarised the situation in the following terms: 'Poor people are usually excluded from the environmental decision-making process, and once a policy is made, they are usually powerless to change it' (3–4).

For South (2010), the depletion of resources caused by environmental degradation can only exacerbate such existing social division between the well off and the poor:

> In a world of increased scarcity, such inequalities will simply be embedded further and we will face the threat of new social constructions of hierarchies and needs emerging. The resulting competition for resources is likely to produce discrimination and violence based on ethnic, gender and other well-established sources of 'difference' and discrimination. (2010: 237)

Certainly, South's contention that environmental degradation leads to increased division on grounds of ethnicity is well documented in the literature, to the extent that it has been called 'environmental racism' (Spencer et al., 2011). Economic theories as to why exposure to environmental harms apparently varies by race include 'pure discrimination by polluters or politicians in siting decisions; differences in willingness to pay for environmental amenities linked to income or education levels; and variations in the propensity of communities to engage in collective action to oppose the location of potential polluters' (Hamilton, 1995: 1).

Paulido (1996) provides a good overview of the established links between sites of environmental degradation (especially the dumping of toxic materials) and black communities within the United States. Similar results have been replicated at the international level (Alston and Brown, 1993). Boer, Pastor, Sadd, and Snyder (1997) confirm that, statistically, the appearance of discrimination in the location of hazardous waste treatment, storage, and disposal facilities proximate to areas where ethnic minorities live in Los Angeles is not explained by alternative, nonracial factors.

There are also growing data to support the argument that environmental harm disproportionately falls on women. Wachholz (2007) has summarised these effects as being linked to gendered division of labour in many developing countries, where women are disproportionately responsible for subsistence farm labour, child care, care of the sick and infirm, and the gathering of household biofuels and water. Essentially, the effects of environmental degradation on traditional farming industries, water supply, and health increases the workload of these women, 'reducing their opportunities for personal and social development' (Wachholz, 2007: 169), exacerbating the poverty that women the world over disproportionately find themselves subjected to. As the author summarises: 'climate change is likely to actuate gaps between the world's rich and poor, and women are already amongst the poorest' (Wachholz, 2007: 171).

The fact that environmental victimisation falls disproportionately on the marginalised elements of society who lack political sway or power will come as little surprise to most victimologists. It is therefore extremely important that researchers in this area continue to dispel the general impression that environmental degradation as a whole (including climate change), and environmental crime specifically, affects everyone equally. Similarly, criminal justice systems must approach these issues with this in mind. It is to the operation of criminal justice in relation to environmental harm that this chapter next turns attention.

Victims of environmental harm and criminal justice systems

Examples of how criminal justice systems are now being obliged to adapt to meet the challenges of environmental victimisation can be drawn from a number of sources. At a national level, for example, the US Crime Victims' Rights Act 2004 has been applied to victims of environmental harm (Starr, Flack and Foley, 2008; see also *Re: Parker; U.S. v U.S. District Court and W.R. Grace and Co.*).[1] In Europe, the European Union (EU) has adopted Directive 2008/99/EC on the protection of the environment through criminal law. Though the enforcement of environmental legislation by means of criminal law is by no means novel, it has often played a secondary role to administrative sanctions and civil penalties. Indeed, Bell and McGillivray (2008) have drawn on the 'enforcement pyramid' posited by Ayres and Braithwaite (1992) to describe enforcement tactics of many jurisdictions in relation to environmental 'crime'. Essentially this pyramid puts 'persuasion' at its apex, principally because revocation of a company's license to operate is in fact far more damaging than a relatively small fine. The 2008 Directive is thus indicative of a hardening of attitudes over breaches of environmental law, although the response of the EU so far to environmental crime has neglected the impact of such crime on victims (Cardwell, French and Hall, 2011). Potential for greater involvement of such victims can be identified within the earlier 1998 Council of Europe Convention on the Protection of Environment through Criminal Law. Though this treaty has not yet entered into force, and indeed has secured little support, Article 11 provides for the *participation* of environmental groups in relevant criminal proceedings. This has the potential to be a ground-breaking article, raising the possibility of opening up environmental criminal proceedings to wider participation.

At the international level, the most progressive source of legally binding environmental obligations is found in the 1998 United Nations Economic Commission for Europe (UNECE) ('Aarhus') Convention on Access to Information, Public Participation in Decision-Making and Access to Justice in Environmental Matters. The Convention requires governments to bring individuals who may be affected into the decision-making process when environmental issues are at stake. What is significant about the Convention is that, almost uniquely within international law, members of the public are able to refer possible breaches of their rights under the Convention to its Compliance Committee. Of course, whilst the Aarhus Convention offers something of a 'way in' to the international legal order for the individual, in practice this compliance mechanism can be subject to criticism. There is a lack of real compulsive power on behalf of the Compliance Committee to really address victims' complaints and ensure that restitution/apologies from perpetrator states are forthcoming. That said, there is presently a dearth of empirical research on what victims of environmental crime might actually want from a criminal justice (or other) process (Williams, 1997). Consequently, moves to increase the attention paid to such victims in many jurisdictions are (once again) progressing without reference to the victims they are purporting to help (see Rock, 1990).

Discussion

As noted at the start of this chapter, addressing environmental harm through criminal law presents a number of difficulties for most criminal justice systems, even before one begins considering how the *victims* of such harm might be incorporated within that process. What is clear, however, is that whilst more 'traditional' forms of criminal victimisation have over the last few decades come to be recognised and assisted by the world's criminal justice systems, the same cannot be said for victims of environmental harm. I submit, however, that this state of affair does not represent any kind of fundamental incompatibility between environmental harm or its victims and criminal justice, but rather is born from a lack of awareness and the needs to address underlying working cultures throughout the criminal justice process. Both Bell and McGillivray (2008) and Du Rées (2001), for example, argue that the problems inherent in the criminal justice system's approach to environmental crime begin at the stage of investigation and prosecution. Whilst Du Rées puts greater emphasis on the failings of criminal laws or regulations to adequately set the boundaries for the operation of relevant agencies, both Du Rées and Bell and McGillivray effectively agree that the difficulty lies with the working practices of those agencies, rather than with fundamental incompatibilities between environmental concerns and criminal justice *per se*.

Sometimes such restrictive working practices may be deeply enmeshed with wider power interests in society. For example, Ebeku (2003) has discussed concerns that, well into the new century, Nigerian judges were prioritising the country's economic reliance on the oil industry over the protection and restitution of the environment or the ordering of compensation/restitution to individual victims or to communities for the massive environmental harms caused by that industry on the Nigerian Delta. The significance of the differing attitudes taken by judiciaries across different jurisdictions to environmental governance has also been highlighted by Kotzé and Patterson (2009). That said, it is important to acknowledge that a lack of understanding amongst criminal justice actors at all levels has been actively addressed in relation to other kinds of previously 'invisible' victims across many jurisdictions, victims of domestic violence and the friends and family of murder victims being two obvious groups (Rock, 1998). Such adaptation of cultures regarding environmental crime and environmental victims has been called for by the Law Society of England and Wales which, in a statement to the UK House of Commons Environmental Audit Committee (2004: 107) noted:

> It frequently takes some time for an environmental crime to be brought before the court and for any appeals to be completed. Until a case has been concluded, an individual cannot obtain the documents relating to the case to assist with their civil action. On occasions the length of time for the criminal case can bring the limitation period for a personal injury claim into play.

The Law Society added that this state of affairs was 'clearly unacceptable' and that

> Given the relationship between environmental offences and human health and living conditions, it might be appropriate for consideration to be given to providing some mechanism whereby indirect compensation can be awarded to those who have suffered the injury. (107)

This suggests that the criminal courts themselves in sentencing environmental offenders must be prepared to address compensation/restitution to victims much more readily.

Of course, from a practical perspective, the key drawback of the critical school has always been precisely that it does not lend itself to easy answers, or to simple characterisations and definitions of victimhood and victims' needs. It is therefore very difficult to reconcile a critical perspective with the suggestion that environmental victims should receive greater recognition by criminal justice systems, which necessarily have to operate on more certain, predefined categories of harm. Nevertheless, it is equally true that the state and the criminal justice system cannot shirk their responsibilities for harms resulting from activities that in some cases are already recognised as breaches of the criminal law and in other cases are receiving stronger social condemnation than ever before (Marquart-Pyatta et al., 2011). To this end, we have noted above that, in the United States at least, broader victim reform agendas have been interpreted as including environmental victims. In other countries and at the international level, such interpretation will not always be possible. Indeed, the definition of 'victim' used in many jurisdictions is quite restrictive (Hall, 2010). Once again, however, what this example does show is that criminal justice as a concept is not fundamentally incapable of incorporating such victims to a greater extent.

One possible solution to reconciling the tensions between traditional modes of criminal justice and victims of environmental crime is to approach the issue from the perspective of human rights. Indeed, it is clear that, in recent decades, human rights have become a cornerstone of debates going on about traditional crime victims (as well as about criminal justice in general) and, as such, will prove a vital component of any 'green victimology' as well. This is all the more certain, given the transnational nature of many environmental harms and therefore the necessary involvement of the international legal order, under which human rights are at present one of the few ways in which individual victims can seek recognition. Indeed, international criminal law may have a vital role to play in the recognition of such victimisation, particularly given the significant victim provisions found within the Rome Statue of the International Criminal Court (Bottigliero, 2004). Some authors and activists have therefore argued in favour of an international crime of 'ecocide' (Gray, 1996), which could either be incorporated by the International Criminal Court (ICC) or to be tried by a new 'international environmental court' (Murphy, 2000).

One particular theme to such discussion, it is suggested, will be that of 'balance'. As noted by Jackson (1990), traditional victimology too has been dominated by balance rhetoric, chiefly concerning the balancing of rights between victim and offender. The same balancing exercise will need to be addressed by green victimologists as well although, in this case, I submit that there are, if anything, more complex issues at stake. Thus, whilst green victimology must tackle the same concern that the rights of environmental offenders will be compromised by more victim involvement in the justice system (the so-called 'zero-sum game'; Jackson, 2004), we have seen that environmental crimes also raise tensions between the economic needs of the broader community and the state as a whole and smaller groups or individual citizens within those communities. Furthermore, to redress environmental harm for some victims may lead to forced changes in industrial practices, putting other victims out of work (Morss, 1996).[2] Thus, it remains clear that, much more so than for many traditional crimes, the balance to be struck may actually be between one set of victims' rights and those of another group of victims, or potential victims, now or in the future.

Finally, if green victimology is to adopt the language of rights it must also, I submit, address another key issue raised by more traditional branches of victimology: the *enforceability* of such rights. There are definite parallels to be drawn between the fledging recognition of rights for environmental victims and those rights ascribed to more traditional victims in that the enforcement mechanisms attached to these 'rights' remain in most cases markedly underdeveloped and lacking true compulsive authority (Jackson, 2004).

Conclusion

As the full impacts of environmental degradation are becoming better understood, it is likely that criminal justice mechanisms will increasingly be called upon respond to the challenges they entail, which ultimately means recognising the needs of those harmed. It is clear that achieving this will require considerable further research from an interdisciplinary perspective and, perhaps most importantly, will require asking environmental victims themselves what they require from a criminal justice system. In this chapter, I have attempted to give a necessarily brief summation of this research agenda on what is set to become (and for some already is) a key issue for academics in most fields of social inquiry in the twenty-first century.

Notes

1. 09–70529, 09–70533 (9th Cir).
2. Although for counterarguments on this point see Goodstein (1994).

References

Alston, D. and Brown, N. (1993) 'Global threats to people of color', in D. Bullard (ed.) *Confronting Environmental Racism: Voices from the Grassroots*. Cambridge: South End Press, pp. 179–194.

Ayres, I. and Braithwaite, J. (1992) *Responsive Regulation: Transcending the Deregulation Debate*. New York: Oxford University Press.

Bell, S. and McGillivray, D. (2008) *Environmental Law* (7th edn). Oxford: Oxford University Press.

Berrittella, M., Bigano, A., Roson, R. and Tol, R. (2006) 'A general equilibrium analysis of climate change impacts on tourism', *Tourism Management*, 27(5): 913–924.

Boer, J., Pastor, J., Sadd, L. and Snyder, L. (1997) 'Is there environmental racism? The demographics of hazardous waste in Los Angeles County', *Social Science Quarterly*, 78(4): 793–810.

Bottigliero, I. (2004) *Redress for Victims of Crimes under International Law*. Leiden: Martinus Nijhoff Publishers.

British Broadcasting Corporation (2007) *EU Pollution Deaths Cost Billions* [online]. Available at http://news.bbc.co.uk/1/hi/sci/tech/4444191.stm. Accessed on 12 August 2012.

Brook, D. (1998) 'Environmental genocide: native Americans and toxic waste', *American Journal of Economics and Sociology*, 57(1): 105–113. Reprinted in *Environmental Crime: A Reader*, edited by R. White. Cullompton: Willan Publishing, pp. 392–398.

Brown, K., Turner, R., Hameed, H. and Bateman, I. (1997) 'Environmental carrying capacity and tourism development in the Maldives and Nepal', *Environmental Conservation*, 24(4): 316–325.

Brunnèe, J. (1995) 'Environmental security in the twenty-first century: new momentum for the development of international environmental law?' *Fordham International Law Journal*, 18: 1742–1747.

Bryant, B. (1995) *Environmental Justice*. Boston: Island Press.

Čapek, S. (1993) 'The environmental justice frame: a conceptual discussion and an application', *Social Problems*, 40(1): 5–24.

Cardwell, P., French, D. and Hall, M. (2011) 'Tackling environmental crime in the European Union: the case of the missing victim?' *Environmental Liability*, 19(2): 35–44.

Chalecki, E. (2001) 'A new vigilance: identifying and reducing the risks of environmental terrorism', *Global Environmental Politics*, 2(1): 46–64.

Chinese Network Television (2011) China passes amendments to criminal law [online]. Available at http://english.cntv.cn/program/china24/20110226/107569.shtml. Accessed 12 August 2012.

Christie, N. (1986) 'The ideal victim', in E. Fattah (ed.) *From Crime Policy to Victim Policy*. Basingstoke: Macmillan, pp. 17–30.

Cullinan, P., Acquilla, S. and Dhara, V. (1996) 'Long term morbidity in survivors of the 1984 Bhopal gas leak', *National Medical Journal of India*, 9(1): 5–10.

Dobson A. (1998) *Justice and the Environment: Conceptions of Environmental Sustainability and Dimensions of Social Justice*. Oxford: Oxford University Press.

Dolan, P. and Moore, S. (2007) 'From preferences to experiences: valuing the intangible victim costs of crime', *International Review of Victimology*, 14(2): 265–280.

Domroes, M. (2001) 'Conceptualising state-controlled resort islands for an environment-friendly development of tourism: the Maldivian experience', *Singapore Journal of Tropical Geography*, 22(2): 122–137.

Duffy, R. (2010) *Nature Crime: How We're Getting Conservation Wrong*. New Haven: Yale University Press.

Du Rées, H. (2001) 'Can criminal law protect the environment', *Journal of Scandinavian Studies in Criminology and Crime Prevention*, 2: 109–126. Reprinted in: *Environmental Crime: A* Reader, edited by R. White. Cullompton: Willan Publishing, pp. 638–655.

D'Silva, T. (2006) *The Black Box of Bhopal: A Closer Look at the World's Deadliest Industrial Disaster.* Bloomington: Trafford Publishing.

Ebeku, K. (2003) 'Judicial attitudes to redress for oil-related environmental damage in Nigeria', *Review of European Community and International Environmental Law*, 12(2): 199–208.

Elias, R. (1983) *Victims of the System: Crime Victims and Compensation in American Politics and Criminal Justice*. New Brunswick: Transaction.
Elias, R. (1986) *The Politics of Victimization: Victims, Victimology and Human Rights*. New York: Oxford University Press.
Farrall, S. and Maltby, S. (2003) 'The victimisation of probationers', *Howard Journal of Criminal Justice*, 42(1): 32–54.
Garkawe, S. (2004), 'Revisiting the scope of victimology – how broad a discipline should it be?' *International Review of Victimology*, 11: 275–294.
Gibbs, C., Gore, M., McGarrell, E. and Rivers, L. (2010) 'Introducing conservation criminology: towards interdisciplinary scholarship on environmental crimes and risks', *British Journal of Criminology*, 50(1): 124–144.
Giddens, A. (1990) *The Consequences of Modernity*. Cambridge: Polity Press.
Goodstein, E. (1994) *Jobs and the Environment: The Myth of a National Trade-off*. Washington DC: Economic Policy Institute.
Gray, M. (1996) 'The international crime of ecocide', *California Western International Law Journal*, 26: 215–272.
Green, P. and Ward, T. (2004) *State Crime*. London: Pluto Press.
Hall, M. (2010) *Victims and Policy Making: A Comparative Approach*. Cullompton: Willan Publishing.
Hamilton, J. (1995) 'Testing for environmental racism: Prejudice, profits, political power?' *Journal of Policy Analysis and Management*, 14(1): 107–132.
Heckler, H. (2011) *Peace and Sustainable Development through Environmental Security*. The Hague: Institute for Environmental Security.
Hillyard, P. and Toombs, D. (2003) 'Introduction', in P. Hillyard, D. Toombs and C. Pantazis (eds) *Beyond Criminology: Taking Harm Seriously*. London: Pluto Press, pp. 1–9.
Hofrichter, R. (1993) *Toxic Struggles: The Theory and Practice of Environmental Justice*. Philadelphia: New Society Publishers.
Hough, M. (1986) 'Victims of violent crime: findings from the first British crime survey', in E. Fattah (ed.) *From Crime Policy to Victim Policy*. Basingstoke: Macmillan, pp. 117–132.
Hough, P. (2012) *Environmental Security: An Introduction*. Abingdon: Routledge Publishing.
Hulsman, L. (1986) 'Critical criminology and the concept of crime', in H. Bianchi and R. van Swaaningen (eds) *Abolitionism: Towards a Non-Repressive Approach to Crime*. Amsterdam: Free University Press.
International Association for the Study of Insurance Economics (2009) *The Insurance Industry and Climate Change – Contribution to the Global Debate*. Geneva: The Geneva Association.
Ivanic, M. and Martin, W. (2008) 'Implications of higher global food prices for poverty in low-income countries', *World Bank Policy Research Working Paper No. 4594*. Washington: The World Bank.
Jackson, J. (1990) 'Getting criminal justice out of balance', in S. Livingstone and J. Morison (eds) *Law, Society and Change*. Aldershot: Dartmouth, pp. 114–133.
Jackson, J. (2004) 'Putting victims at the heart of criminal justice: the gap between rhetoric and reality', in E. Cape (ed.) *Reconcilable rights? Analysing the Tension between Victims and Defendants*. London: Legal Action Group, pp. 65–80.
Jaggi, B. and Freedman, M. (2006) 'An examination of the impact of pollution performance on economic and market performance: pulp and paper firms', *Journal of Business Finance and Accounting*, 19(5): 673–713.

Jasparro, C. and Taylor, J. (2008) 'Climate change and regional vulnerability to transnational security threats in Southeast Asia', *Geopolitics*, 13(2): 232–256.

Kotzé, L. and Paterson, A. (2009) *The Role of the Judiciary in Environmental Governance: Comparative Perspectives*. Amsterdam: Wolters Kluwer.

Lee, M. (2007) *Human Trafficking*. Cullompton: Willan Publishing.

Lee, T. (2011) 'Gulf Shrimp season begins in Louisiana under Oil-Black cloud' [online], *The Huffington Post*. Available at http://www.huffingtonpost.com/2011/08/22/start-of-louisianasshrim_ n_933472.html?ir=Green&ncid=edlinkusaolp00000008. Accessed 26 June 2012.

Lee, T-P. (2009) *A Welfare Approach to Mitigating Environmental Injustice: Exploring Needs of Pollution Victims* [online]. Available at http://www.welfareacademy.org/pubs/international/epckdi/2.pdf. Accessed 26 June 2012.

Lee, M. and Blanchard, T. (2010) *Health Impacts of Deepwater Horizon Oil Disaster on Coastal Louisiana Residents*. Baton Rouge: Louisiana State University Press.

Lobell, D., Schlenker, W. and Costa-Roberts, J. (2011) 'Climate trends and global crop production Since 1980', *Science*, 5 May.

Lynch, M. and Stretesky, P. (2001), 'Toxic crimes: examining corporate victimization of the general public employing medical and epidemiological evidence', *Critical Criminology*, 10: 153–172.

Macleod, A., Pautasso, M., Jeger, M. and Haines-Young, R. (2010) 'Evolution of the international regulation of plant pests and challenges for future plant health', *Food Security*, 2: 49–70.

Mandalia, S. (2005) 'The spice traders' tale', *New Statesman*, 7 March: 14.

Mares, D. (2010) 'Criminalizing ecological harm: crimes against carrying capacity and the criminalization of eco-sinners', *Critical Criminology*, 18: 279–293.

Marmo, M. and La Forgia, R. (2008) 'Inclusive national governance and trafficked women in Australia: otherness and local demand', *Asian Criminology*, 3: 173–191.

Marquart-Pyatta, S., Shwomb, R., Dietza, T., Dunlapc, R., Kaplowitza, S., McCrighta, A. and Zahrand, S. (2011) *Understanding Public Opinion on Climate Change: A Call for Research, Environment: Science and Policy for Sustainable Development*, 53(4): 38–42.

McBarnet, D. (1983) 'Victim in the witness box – confronting victimology's stereotype', *Crime, Law and Social Change*, 7: 293–303.

McMahon, M. (1990) '"NET-WIDENING" vagaries in the use of a concept', *British Journal of Criminology*, 30(2): 121–149.

Miers, D. (1997) *State Compensation for Criminal Injuries*. London: Blackstone.

Mörner, N-A., Tooley, M. and Possnert, G. (2004) 'New perspectives for the future of the Maldives', *Global and Planetary Change*, 40(1–2): 177–182.

Morss, E. (1996) 'Clean air act implementation: an industry perspective', *Pace Environmental Law Review*, 14(1): 63–68.

Mullier, E. (2010) 'The emergence of criminal competence to enforce EC environmental law: Directive 2008/99 in the context of the case-law of the European Court of Justice', *Cambridge Student Law Review*, 6(1): 94–116.

Murphy, S. (2000) 'Does the world need an international environmental court?' *George Washington Journal of International Law and Economics*, 32: 333–349.

Murray, S. (2006) 'Poverty and health', *Canadian Medical Association Journal*, 174(7): 923.

Mythen, G. and Walklate, S. (2006) Criminology and terrorism: which thesis? Risk society or governmentality? *British Journal of Criminology*, 46(3): 379–398.

Nemani, R., White, M., Cayan, D., Jones, G., Running, S., Coughlan, J. and Peterson, D. (2001) 'Asymmetric warming over coastal California and its impact on the premium wine industry', *Climate Research*, 19: 25–34.

Patz, J., McGeehin, M., Bernard, S., Ebi, K., Epstein, P., Grambsch, A., Gubler, D., Reiter, P., Romieu, I., Rose, J., Samet, J. and Trtanf, J. (2000) 'The potential health impacts of climate variability and change for the United States: executive summary of the report of the health sector of the U.S. national assessment', *Environmental Health Perspectives*, 108(4): 367–376.

Pepper, D. (1993) *Eco-Socialism: From Deep Ecology to Social Justice.* Abingdon: Routledge Publishing.

Pinstrup-Andersen, P. (2009) 'Food security: definition and measurement', *Food Security*, 1: 5–7.

Pointing, J. and Maguire, M. (1988) 'Introduction: the rediscovery of the crime victim', in M. Maguire and J. Pointing (eds) *Victims of Crime: A New Deal?* Milton Keynes: Open University Press, pp. 1–13.

Possnert, G., Tooley, M. and Mörner, N-A. (2004) 'New perspectives for the future of the Maldives', *Global and Planetary Change*, 40: 177–182.

Reid, H., Sahlén, L., MacGregor, J. and Stage, J. (2007) *The Economic Impact of Climate Change in Namibia: How Climate Change Will Affect the Contribution of Namibia's Natural Resources to Its Economy.* London: International Institute for Environment and Development.

Rock, P. (1990) *Helping Victims of Crime: The Home Office and the Rise of Victim Support in England and Wales.* Oxford: Oxford University Press.

Rock, P. (1998) *After Homicide: Practical and Political Responses to Bereavement.* Oxford: Clarendon Press.

Ruggiero, V. and South, N. (2010) 'Critical criminology and crimes against the environment', *Critical Criminology*, 18: 245–250.

Shelley, L. (2007) 'Human trafficking as a form of transnational crime', in M. Lee (ed.) *Human trafficking.* Cullompton: Willan Publishing, pp. 116–137.

Skinnider, E. (2011) *Victims of Environmental Crime – Mapping the Issues.* Vancouver: The International Centre for Criminal Law Reform and Criminal Justice Policy.

Smith, P. (2007) 'Climate change, mass migration and the military response', *Orbis*, 51(4): 617–633.

South, N. (2010) 'The ecocidal tendencies of late modernity: polnational crime, social exclusion, victims and rights', in R. White (ed.) *Global Environmental Harm: Criminological Perspectives.* Cullompton: Willan Publishing, pp. 236–240.

Spencer, M., Garratt, A., Hockman, E., Bryant, B. and Kohn-Wood, L. (2011) 'Environmental justice and the well-being of poor children of color', *Communities, Neighborhoods and Health*, 1(3): 219–233.

Starr, J., Flack, B. and Foley, A. (2008) 'New intersection: environmental crimes and victims' Rights', *Natural Resources and Environment*, 24: 41–52.

Steinberg, T. (2008) *Down to Earth: Nature's Role in American History.* Oxford: Oxford University Press.

Takemura, N. (2007) '"Criticality of environmental crises" and prospect of "complexity green criminology"', *Toxin University of Yokohama Research Bulletin*, 17. Yokohama: Toxin University of Yokohama.

Thompson, E. (1991) *Customs in Common.* London: Penguin Books.

Ullman, R. (1983) 'Redefining security', *International Security*, 8(1): 129–153.

United Nations Environment Programme (2011) *UNEP Ogoniland Oil Assessment Reveals Extent of Environmental Contamination and Threats to Human Health* [online]. Available at http://www.unep.org/newscentre/Default.aspx?DocumentID=2649&ArticleID=8827&l=en. Accessed 26 June 2012.

United Nations Food and Agriculture Organization (2010) *Special Programme for Food Security* [online]. Available at www.fao.org/spfs. Accessed 10 August 2012.

Wachholz, S. (2007) '"At risk": climate change and its bearing on women's vulnerability to male violence', in P. Beirne and N. South (eds) *Issues in Green Criminology: Confronting Harms against Environments, Humanity and Other Animals*. Cullompton; Willan, pp. 161–185.

Warner, K., Afifi, T., Dun, O., Stal, M. and Schmidl, S. (2008) *Human Security, Climate Change and Environmentally Induced Migration*. Bonn: United Nations University – Institute for Environment and Human Security.

Watts, H. (2010) 'The health benefits of tackling climate change', *The Lancet Series on Climate Change*. Available at http://www.wellcome.ac.uk/stellent/groups/corporatesite/@policy_communications/doc uments/web_document/wtx057673.pdf. Accessed 10 August 2012.

Wheatley, M. (1997) 'Social and cultural impacts of mercury pollution on aboriginal peoples in Canada', *Water, Air and Soil Pollution*, 97(1–2): 85–90.

White, R. (2008) *Crimes against Nature: Environmental Criminology and Ecological Justice*. Cullompton: Willan Publishing.

White, R. (2011) *Transnational Environmental Crime: Towards an Eco-Global Criminology*. Abingdon: Routledge Publishing.

Williams, C. (1996) 'An environmental victimology', *Social Science*, 23(1): 16–40. Reprinted in R. White (2009) *Environmental Crime: A Reader*. Cullompton: Willan Publishing, pp. 200–222.

World Health Organization (2006), *Health effects of the Chernobyl accident: an overview*. Available at http://www.who.int/ionizing_radiation/chernobyl/backgrounder/en/index.html. Accessed 26 June 2012.

Young, M. (1997) 'Ideological trends within the victims' movement: an international perspective', in R. Davis, A. Lurigio and W. Skogan (eds) *Victims of Crime* (2nd edn). Thousand Oaks: Sage Publications, pp. 115–126.

Yun, K., Lurie, M. and Hyde, P. (2010) 'Moving mental health into the disaster – preparedness spotlight', *The New England Journal of Medicine*, 263: 1193–1195.

12

'A Gift from the Tropics to the World': Power, Harm, and Palm Oil

Hanneke Mol

Introduction

In the context of a professed energy crisis and a climate crisis spiralling out of control, highly biodiverse natural and sociocultural environments are giving way to vast monocultures of presumably carbon-neutral energy crops. One such crop is the African oil palm (*Elaeis guineensis*).[1] In Colombia, the largest producer of palm oil in the Americas, the official narrative of the National Federation of Oil Palm Growers describes 'the world of the oil palm' as

> A warm and humid one, of shades of green, inhabited by hundreds of animal and plant species. But it is also a world of human relations and labour, where the rural and the urban, the national and the international, agricultural, extractive and industrial activities meet. It is a world where diverse and complementary endeavours merge to form a chain of production, generating wealth and fomenting social development. (Fedepalma, 2006: 2)

This image of palm oil as a social and ecological crop is consistent with the global depiction of biofuels – or, more aptly, of *agro*fuels – as social, environmental, and economically beneficent ways of greening patterns of (excessive) energy consumption. Yet these liquid fuels derived from plant biomass are mired in paradoxes. This chapter, as such, has its basis in ongoing research into the social and environmental harms associated with palm oil production in Colombia's South Pacific region. Here, these harms will be considered through a discussion of power vis-à-vis harm by attending to the human–nonhuman and colonial power relations that mark this industry as well as the agrofuels industry more generally.

The ways human beings relate to, act upon, and interact with the nonhuman[2] are central to green criminological perspectives (see, for example, Halsey and White, 1998; Cazaux, 1999). As such, green criminology has a crucial role to play in extending and deepening the central focus on power relations within critical criminology by turning the lens to the nonhuman, drawing attention to the intricate connection between the human and nonhuman realm, and laying bare mechanisms of ordering *human* life through perspectives on and interventions in the *natural world*. The principal argument put forward in this chapter is that colonial forms of control, appropriation, and territorial ordering are intimately related to the power relation between the human and the nonhuman. Accordingly, I will illustrate through the case of Colombian palm oil, that the notion of coloniality[3] can advance criminological debate on the working of power and harm in contexts like these.

The social and environmental promise of agrofuels

In the framework of increasing concern over diminishing crude oil reserves, rising oil prices, and dependence on oil supplies from unstable production regions, coupled with demands to reduce greenhouse gas emissions to mitigate climate change, agrofuels are part of a rapidly expanding global market and figure centrally in discussions within and among the spheres of politics, academia, and civil society. In the official discourse of state, corporate, and regional through to supranational official bodies, agrofuels are promoted as socially beneficent green alternatives, believed to be capable of steering the economy away from fossil fuel dependence. Without requiring any fundamental changes to current production and consumption patterns of energy, projected levels of economic growth are left unchallenged (Dauvergne and Neville, 2009). A developmental promise moreover posits biofuel production as an opportunity to foster economic growth, reduce poverty levels, and improve social welfare in regions of production in the global South (Kessler et al., 2007; Levidow and Paul, 2010).

However, in reality this green dream scenario is met with increased contestation and controversy (Franco et al., 2010), signified amongst other things by terminological juxtaposition of 'biofuels' in the discourse of the industry proponents versus 'agrofuels' in the discourse of the critic. According to its critics, the energy crops in question are linked to processes that *undermine* life rather than contributing to life processes, as the prefix 'bio' would imply (McMichael, 2009; Cerdas Vega, 2009). In sharp contrast to the agrofuels fairy-tale, there is a different, darker story to be told, in which criminology, too, must make itself heard. This is the bleak story of the sheer ecological, social, and cultural destruction waged through the production and consumption of this 'gift' from the tropics to the world.

This chapter aims to shed critical light on the green and social promise of the 'agrofuels solution' that generates an array of harms against the people, plant and animal species, and ecosystems of the production regions in question. I will first provide an introductory overview of the ecological dimension of the implied harms. This will be followed by the associated social and cultural consequences for plantation workers and local inhabitants of zones of agrofuels and, more specifically, zones of palm oil production. Lest there be any doubt, this should not be interpreted as a separate, isolated account of ecological versus social issues; there is an interwovenness that is accorded principal attention in green criminological analyses (for example White, 2002; Halsey, 1997).

The reality of agrofuels

The environmental impact of agrofuel production to a certain extent is linked to the type of feedstock[4] that is cultivated. In the case of palm oil production it is not so much the tree itself that is the problem as its cultivation in industrial monocultures; vast plantations made up of orderly rows of a single species. It is this mono-crop cultivation model that characterises the production of agrofuels globally (Cerdas Vega, 2009). Although to the untrained eye monocultures may look like forests, in fact they bear no resemblance to the complexity, resiliency, and species diversity of forest ecosystems (Altieri, 2009). The imposition of uniformity and simplification that monocultures impose on ecosystems not only intends to maximise yields but moreover facilitates processes of efficient and thorough management and control (see Scott, 1998) of both natural processes and of human life.[5] Although by no means an exclusively capitalist phenomenon, the imperative to continuously increase productivity *intrinsic* to the logic of capitalist accumulation gives rise to cycles of agro-ecological restructuring that entail the radical simplification of and ever more intensified exploitation of nature (Moore, 2000).

Favourable soil conditions and climatic requirements have it that, in the case of an encroaching agrofuels industry, such processes of uniformity, manipulation, and exploitation are to a great extent accompanied by the clearance of vast areas of highly biodiverse moist tropical forests. In the Colombian Pacific, too, palm oil is a major driver of deforestation (Pérez-Rincón, 2008). The plantations that replace these forests accommodate extremely low levels of biodiversity, entailing a significant and largely irreversible decline in endemic flora and fauna species (Fitzherbert et al., 2008; Wilcove and Koh, 2010).[6] To the loss of biodiversity must be added the impact of soil erosion and the substantial quantities of carbon released into the atmosphere via processes of deforestation and land-use change (Smolker et al., 2008). Taking into consideration that tropical ecosystems account for roughly 46 per cent of the world's carbon stored in the terrestrial biosphere (Danielsen et al., 2008), land conversion geared towards agrofuel production in effect amounts to an actual *increase* in net greenhouse gas emissions.

Moreover, and adding to the paradoxical reality of agrofuels, a manifest dependence on the use of fossil fuels *throughout* the chain of agrofuel production, from cultivation through to processing and transport (Smolker et al., 2008), speaks against a proclaimed 'shift away' from fossil fuel dependence. Owing to ecosystem simplification and the exhaustion of soil nutrients, monocultures are, moreover, vulnerable to disease and pests problems. Intrinsic to the agro-industrial mono-crop model, then, are the huge amounts of pesticides, herbicides, and fertilisers required. These agrichemicals, too, often have their basis in fossil fuels. Together with runoff and wastes from processing plants, large amounts of liquid and solid waste as such contaminate soils, rivers, and groundwater (McCarthy and Zen, 2010). This affects fish populations, nearby flora and fauna, and also forms a direct and indirect threat to the health of local inhabitants and plantation workers. To the pollution of waterways must be added the *depletion* of water sources; monocultures place excessive demands on water supplies and hence are linked to water shortages and hydrological imbalances (Altieri, 2009; Smolker et al., 2008).

The impacts of deforestation, soil erosion, the contamination of rivers, and the depletion of water sources are experienced with alarming intensity in the Colombian Pacific. It is also worthwhile pointing to the microclimatic changes brought about in this region, entailing the deregulation of precipitation regimes, alteration of wind patterns, and rising temperatures in areas immediately surrounding the oil palm plantations (Vélez Torres, 2010: 83–85). By contrast, the Colombian National Federation of Oil Palm Growers maintains that mature oil palm plantations are 'true forest landscapes that accommodate numerous flora and fauna species', said to 'create microclimates and environments favourable to the sustainability of the crops and well-being of surrounding populations' (Fedepalma, 2006: 19).

Hence, the oil palm is depicted as an ecological *and* a social crop. From the above statements about ecological impacts, interconnected as they are with bodily, social, and cultural dimensions of existence, it follows that the proclaimed social character of palm oil, and of agrofuels more generally, is hard to maintain.[7] A main critique of agrofuels concerns the threat posed to food sovereignty and security, due to the fact that the production of agrofuels actively competes with the production of food. The energy crops are linked to food shortages and sharp rises in prices of basic staple foods such as maize (with an estimated rise that could reach 41 per cent by 2020), wheat (30 per cent by 2020), and vegetable oils (76 per cent by 2020), impeding the economic access of many often already marginalised people to sufficient food (Ziegler, 2007).[8]

In Colombia, embedded in the broader context of an export-oriented agribusiness, the expansion of oil palm cultivation undermines, disrupts, and displaces local production and subsistence practices. As a consequence, communities are no longer able to hold on to their local models of cultivation, characterised by a diversification of cultivated crops interspersed with areas of naturally growing plant species and responding to the material and social reproduction of its populace (Restrepo, 2005; Ramírez Vidal, 2007). As the oil palm competes with other crops for space and depletes water sources, the cultivation of traditional crops geared towards local subsistence patterns and needs will be seriously debilitated. Moreover, the contamination of rivers results in substantial declines in fish populations, whilst losses in flora and fauna due to habitat destruction affect hunting activities (Vélez Torres, 2010). Together, these factors destroy the fundamental bases of local dietary patterns, food security, and economic self-sufficiency. In the Pacific coast region that stands out for its natural abundance, the levels of social and economic inequality and exclusion are staggering, with rates of chronic malnutrition and poverty well exceeding the national average (Flórez López and Millán Echeverría, 2007). Here, palm oil is produced in a context of armed conflict and furthermore is part of 'a model of development that fails to benefit the inhabitants of this region' (ibid.: 25).

As Levidow and Paul (2010) have stated more generally, the conversion of lands previously geared to subsistence and/or local agricultural systems results in small-holders and local communities being incorporated into or effectively forced into the global commodity market. As nonintensive land-use practices do not fit capitalist accumulation scripts of productivity, rehabilitating or putting 'under-utilised' lands to 'proper' use comes to be presented as a beneficial and crucial developmental intervention (Restrepo, 2005). The consequent land-use change, concentrated land ownership, and extended reach of private property regimes frequently disarticulate existing social and cultural relations and practices. This, in turn, has far-reaching consequences for territorial and cultural integrity, quality of life, and human dignity.

It is crucial to note how access to, the use of, and *unique* ways of using territory are fundamental for the physical and cultural survival of communities that strongly relate to and depend on the socio-natural environments of the lands they inhabit, often ancestrally. As documented vividly by Escobar (2008), the centrality of the forest and aquatic environments to many of the Afro-Colombian communities inhabiting the Pacific Coast region of Colombia is an illustrative case in point. Socio-territorial affiliation is strongly linked to the region's extensive river network as a source of income and sustenance; it is an infrastructure central for transport, settlement and communication patterns and a cultural reference point for identity formation and sense of belonging (Escobar, 2008; Ventes et al., 2008).

Yet access to rivers is ever more restricted or is considered too dangerous due to the presence of armed actors (Ventes et al., 2008), and patterns of dispersed settlements along rivers are under increased pressure from an expanding palm oil industry. As argued by Escobar (2008), these processes can be interpreted as part of a broader push towards the 'reconfiguration of the biophysical and cultural landscapes of the Pacific' (64) that seeks to erase the economic, cultural, and ecological characteristics that are definitive of the ways of being of many Afro-Colombians inhabiting this region. In this context, communities suffer not only forced *dis*placement but situations of *em*placement, too; the restriction of mobility and routine spatial practice (Escobar, 2008). This is an experienced 'symbolic and psychological displacement' (Ventes et al., 2008) that disrupts existing ways and dynamics of relating to the territory and the right to free movement. In effect, those that refuse to leave their lands become prisoners in their own territories, unable to enjoy the fruits of what by legal title and ancestrally belong to them.

Despite examples to the contrary, all too often the imposition of oil palm cultivation is part of a context that brings destruction to and dramatically undermines the integrity and existence of human and nonhuman life. Both are seemingly expendable in the service of a lucrative market and, let's not forget, of the lifestyles of those in the affluent West, including many a reader of and contributor to this book.[9]

Green criminology, power, harm, and the critical tradition

The analysis and theorisation of the harms in question call for profound interrogation of the type of power relations at work in this context. A power-based approach to harm can deepen the debate on the conceptualisation of harm within criminology and is indispensable in exposing and dissolving the organising principles that have something to say about where harm is perpetrated; against whom and what; the nexus and shifting boundaries between crime and harm; and who decides on the terms of the debate – as regards the terminology deployed as well as who is granted a subject position to speak from (for this latter aspect, see Mignolo, 2000). I suggest that we reflect upon the production of harm vis-à-vis power differentials operating along human–nonhuman lines of differentiation *as intimately related to* colonial technologies of power and logics of ordering.[10]

The approach I propose here fits neatly into evolving critical and globally oriented perspectives within green criminology that have taken up the crucial need to incorporate environmental issues into their analyses of the inequitable, exploitative, and oppressive outcomes of power differentials. Lynch's call upon critical criminologists, first explicitly articulated over two decades ago, was premised on the recognition of 'the ability of powerful groups to manipulate and use race, class, gender *and the environment* to preserve the basis of their power' (Lynch, 1990: 1; emphasis added).

By interrogating existing power relations, green criminological analysis embeds itself in, continues, and *extends and deepens* the critical criminological tradition by taking the analysis into the socio-ecological realm. How power operates through green issues is interlinked with the ordering of an array of social relations. Having considered this, in the following I take up, and limit myself to, the place accorded within green criminological perspectives to reflection upon human–nonhuman and colonial power relations in perspectives on harm. Subsequently this will lead me back to the cultivation oil palm in the Colombian Pacific in somewhat additional detail.

Human and nonhuman relations and interaction

Drawing on eco-philosophical principles, Halsey and White (1998; see also White, 2007) problematise the anthropocentric attitudes to the human-environment nexus that are the basis of much ecological destruction and, intertwined with this, of much social and cultural destruction. From a perceived discontinuity between the human and natural realm, anthropocentric notions are constructed upon a set of dualisms that externalise, inferiorise, and objectify nonhuman nature. Cazaux (1999), in an instructive overview, traces the history and implications of the entrenched anthropocentrism that as such inheres in the ways human beings relate to the rest of nature and animal species. Criminology is not exempted from the critique thus voiced, which leads, Cazaux (1999) to stress the need for criminologists to analyse harm from a 'non-speciesist angle'.

It is generally from such a position that green criminologists argue for the necessity to confront the hegemony of ways of social organisation that have their basis in an economic-reductionist rationality and instrumentalism consistent with which only those environmental issues that interfere directly with sectional interests and narrow conceptions of *human* well-being and needs are deemed problematic (Halsey and White, 1998; Walters, 2010). By such accounts, nature, as human property and 'natural capital' at our disposal, is reduced to an exterior realm to be exploited and appropriated. Accordingly, ecological considerations are largely absent; the economic and political dictates and cultural logic of global capitalism go unquestioned, sustaining what are in actual fact unsustainable and unjust social and ecological relations (Halsey, 1997; White, 2002). Proposed solutions to what are by implication a limited range of environmental concerns are sought through the market (more market), technological innovation, and individualised forms of responsible behavior (recycle, buy *green* – not less!).

In view of the 'human exceptionalism' that thus permeates human-nature relations and interactions, Beirne's (1999) account of the various reasons for criminologists to turn their lenses to animal abuse raises broader questions about the persistence and social acceptance of (or, indifference to) human-inflicted suffering upon animals and the destruction of the natural environment. Why are so many of these acts scarcely perceived as harmful or abusive, let alone criminal, and consequently – in either form – perceived as unworthy of criminological attention? These are questions that ought to be at the centre of criminology, not only to establish a more comprehensive analysis of the operation of power, but also as a necessary corrective to disembodied and disembedded anthropocentric understandings of harm that stem from criminological neglect of nonhuman well-being and neglect of the inextricable connection among ecological issues, bodily integrity, social and cultural embedding, and human well-being (Halsey, 1997; Benton, 1998).

In addressing these questions it is important to note that the human exceptionalism that reaches a zenith in the modern, rationalist tradition is embedded in a cultural complex shot through with a colonial logic of societal ordering (Quijano, 2007). Along these lines, the argument advanced here rests upon the view that processes of ordering and exploiting nature, and with that the ordering of human life through perspectives on and intervention in the natural world, are significantly imbued with a colonial dimension. The question that thus emerges is to what extent the colonial figures in green criminology.

Colonial power relations; conquering and exploiting 'peripheral' natures and peoples

The intersection between different systems of social stratification as well as mechanisms of 'othering' that cross the species boundary are central in green criminology. Hence Beirne and South (2007: xx) identify 'gender inequalities, racism, dominionism and speciesism, classicism, the north/south divide, the accountability of science, and the ethics of global capitalist expansion' as 'key concepts for a green criminology'.

Increased emphasis is thereby placed on the fact that a critical green criminology must assume the form of a global criminology (White, 2011), centred on notions of scale and on the cross-boundary operation of 'the politics of power, harm and justice' (Walters, 2010: 314). This directs attention to the analysis of relations of production and consumption that, conducive to or more suitable to the service of the needs of capital (White, 2002), compel the production *and externalisation* of environmental harm. Examining local–global encounters and connections is crucial in themselves, illustrated, for instance, by Boekhout van Solinge (2010), who draws attention to the connection between processes of deforestation in the Amazon rainforest and patterns of unsustainable meat consumption in Western societies.

In keeping with such observations, an encompassing analysis of harm in the context of palm oil production must attend to the driving forces behind this industry. The connection between the ambitious biofuel targets of the European Union and United States and the creation of incentives for the conversion of land elsewhere is one such aspect that cannot be ignored in this respect.[11] It is crucial to reflect upon the colonial power mechanisms at play. This applies not only to the North–South divide, but also to centre-periphery relations *within* regions of the global South that assign a peripheral status of existence to certain regions, consistent with a view of the environments and human inhabitants of these regions as instrumental, manipulable, and to an extent expendable, even in the process of implanting dominant models of development and/or industries geared to support the lifestyles of those in the centre/North. Conquering these lands and the minds of the people inhabiting these lands to exploit their ecological and cultural diversity is either considered unproblematic or altogether beyond consideration.[12]

Within green criminology, South (2007), amongst others, has advocated the need for a postcolonial perspective that attends to the colonial expropriation of the knowledge and nature of the developing world through the patenting of genetic properties of plant, animal, or human biological material; South refers to such acts of bio-piracy as 'an extension of colonial exploitation into late modernity' (2007: 241). These are typical practices of the exertion of control over, exploitation of, and manipulation of the people and the natural environments of the periphery by the Western world with absolute disregard for the social, cultural, and ecological practices and relations disrupted as a result. This is, moreover, accompanied by a persistent 'exclusion from discourse' (South, 2007). A propensity to render nonscientific, non-Western ways of thinking, seeing, and being invisible, or to cast such systems of thought off as mere folklore or the cosmovisions or belief systems of cultures, reduces these to *objects* of study, rather than being acknowledged as valid forms and producers of knowledge in their own right (Tuhiwai Smith, 1999; Walsh, 2007).

In working towards an eco-global perspective in criminology, White (2011) takes up the methodological and theoretical implications of the spatial dimension and local–global dynamics of environmental issues for the analysis of harm. Considerations of place, scale, and the transference of harm are thereby amongst others tied to a focus on the production of knowledge, hence encompassing social, ecological, and epistemological reflection. That is to say, attention is called to the fact that the world is viewed and perceived differently through different eyes and from different localities, histories, and subjectivities. For these to be perceived and accordingly allowed into the analysis, criminologists must confront the 'hegemony of the centre' (White, 2011: 29) and the corresponding marginalisation of 'voices from below' in academic writing and research:

> To speak of the transnational…demands an appreciation that the 'transnational' is very often conceptually located within familiar scholarly universes. The development of a truly global criminology

will require breaking the chains of parochialism, elitism, and (implicitly) [of] a colonialist mentality. (White, 2011: 29)

From these examples it follows that a focus on local–global/centre-periphery dynamics in the analysis of harms against human beings and nonhuman nature in recent and leading green criminological works demonstrates a critical awareness of the enduring legacy of colonial forms of control, domination, and exploitation. These contributions notwithstanding, overall, the manifestation of a colonial dimension of power and logic of ordering in the production, transference, and analysis of harm warrants more explicit and comprehensive reflection and articulation within (green) criminology. It is instructive to turn in somewhat more detail to the case of the Colombian Pacific to illustrate this.

Palm oil production in the Colombian Pacific

In 2010, the total area planted with oil palms in Colombia reached 404,104 hectares,[13] which translates into an 11.5 per cent increase in comparison to the previous year. Principal destinations for exports to Europe are the United Kingdom and the Netherlands, which in 2010 imported 10.6 per cent and 5.5 per cent of Colombian palm oil, and 12.3 per cent and 22.9 per cent of Colombian palm kernel oil, respectively (Fedepalma, 2011).[14]

The Pacific Coast region of Colombia, considered one of the most biodiverse regions in the world, is inhabited predominantly by Afro-Colombian communities, descendants from enslaved Africans who constitute over 90 per cent of the population in this region (Oslender, 2008; Escobar, 2008). Indigenous groups of various ethnic descents make up roughly 6 per cent of the population, the remaining 4 per cent being of mixed race (Flórez López and Millán Echeverría, 2007). The heterogeneity of this region, in terms of its cultural and ethnic diversity, forms of sociocultural organisation, its 'disorderly' nature and local ways of relating to the natural environment (Escobar, 2008), have come under increasing pressure from the homogenising mechanisms of dominant models of development, sustained to a large extent by the 'armed machinery' that operates in the region and that is subjecting many of its inhabitants to a new diaspora (Arocha Rodríguez and Moreno Tovar, 2007). It is imperative to view the imposition of oil palm cultivation onto this region as yet another episode in a historical *continuum* of violence, subjugation, displacement, and emplacement lived by Afro-Colombians (Mosquera Rosero-Labbé, 2007).[15]

The transformation of complex and diverse biophysical processes and cultural and socio-ecological relations in regions targeted for agrofuel cultivation and production is consistent with what Scott (1998) has referred to more generally as the delimitation of a particular field of vision and associated processes of the material transformation of 'disorderly' ecological, social, and cultural realities for purposes of control, management, and manipulation. The lands and territories thus laid claim to moreover become the object of an 'international division of nature', entailing new processes of appropriation and territorial ordering, redistribution of ownership of nature, and geopolitical reconfiguration of the distribution of nature's goods (Vélez and Vélez, 2008). As Alimonda (2011) has argued more broadly:

> A persistent coloniality impacts on Latin American nature. The latter, both as biophysical reality (its flora, fauna, human inhabitants, the biodiversity of its ecosystems) and its territorial configuration (the sociocultural dynamic that significantly articulates these ecosystems and landscapes) is depicted, in global hegemonic thought and by local elites, as a subaltern space open to exploitation, devastation, and reconfiguration according to the demands of current regimes of accumulation. (Alimonda, 2011: 22, my translation)

It is crucial to note that such processes of colonisation are imposed from the interior, too (Vélez and Vélez, 2008). This point is frequently reiterated in works that address the situation in the Colombian Pacific. In view of the push towards oil palm in the Pacific region, an often-heard critique is that, rather than respecting the autonomy and capacity of local populations to decide on their own forms of development, 'from the centre of the country it is decided what is, or is not, in the interest of this region' (Arboleda Montaño, 2008: 114).

At this point it must be mentioned that the Afro-Pacific is a region that historically has been portrayed as backward, and its inhabitants have been portrayed as racially and culturally inferior. An imagined geographical and racial hierarchy, referred to as 'andinocentrism', has created a centre-periphery relation that relegates the territories and the people of the Pacific coast region to an inferior status (Arocha Rodríguez and Moreno Tovar, 2007). In this model, the Andes are conceived as the locus of civilisation, from which progress must spread out to places as yet not (fully) civilised and developed. The 'whitening' (*blanqueamiento*) of the region through processes of colonisation and mestisation as such constitutes a trajectory towards heightened development and compliance with dominant ideas of 'progress' (Arocha Rodríguez and Moreno Tovar, 2007).

Representations of the Colombian Pacific along lines of an andinocentric imaginary are not something of the past though, but effectively in place today.[16] The push towards an agro-industrial production model of palm oil is considered consistent with an andinocentric or Euro-Andean model (Escobar, 2008) of 'development' and the promotion of homogeneities. In the light of this, Ramírez Vidal (2007), looking back upon his observations in one palm oil ridden region in the southern Colombian Pacific, asserts:

> The oil palm corporations had become the new symbol of mestizo and white cultural penetration. The control of Afrocolombian groups over their environs was being seriously threatened and undergoing profound socioeconomic and cultural transformation. (410)

Critics stress how the palm oil industry has reinvigorated exploitative configurations of humans and nature that resonate with a colonial past, even to the extent of subjugating the region's inhabitants to a new form of slavery by an imposed alteration of the use of the lands and the disruption of personal relations and family dynamics (Ramírez Vidal, 2007). Escobar, too, has characterised the oil palm industry – from its introduction into Colombia and the reconfiguration of human and nonhuman relations to the homogenising discipline of the plantation and the management, categorisation, and manipulation of nature and social relations – as 'infused with colonial overtones' (Escobar, 2008: 70).

The gift of palm oil to the world leaves the people and the environments of the tropics with contaminated soils, groundwater, and rivers; habitat destruction; ecosystem disturbances; the loss of flora and aquatic and animal species; and processes of displacement and emplacement that inflict a whole range of physical, psychological, social, and cultural consequences upon local communities. The gift of (human and nonhuman) life *itself* can seemingly be dispensed with, on a scale and with a level of severity unparalleled by numerous matters of 'traditional' criminological concern. If anything, these dynamics and processes, this *destruction* of life in its many forms and qualities, demand the attention of a critical criminology concerned with the operation of power and questions of 'harm'.

By way of conclusion, and where next?

This chapter has questioned the proclaimed social and ecological character of the palm oil industry by detailing the array of adverse impacts that the cultivation and production of these supposedly 'green' and 'socially beneficent' crops inflict upon the people and the ecosystems of the production regions in the global South. Embedded within a critical and globally oriented green criminological perspective, the discussion of harm in the context of palm oil production in South Pacific region of Colombia has revolved around the question of how power operates through green issues. The central argument thus put forward is that the analysis of green issues ought to pay more explicit attention to the colonial aspect; that is to say, in the context here addressed, that power differentials operating along human–nonhuman lines of differentiation are intimately related to a colonial dimension of power. Thereby I have argued that a power-based approach to harm can advance understanding of the organising principles that have something to say about where harm is perpetrated; against whom and what; the nexus and shifting boundaries between crime and harm; as well as who decides on the terms of the debate.

However, in terms of taking this debate further, hence the title 'Where next?' above, it is crucial that this situation is not perceived as one in which local inhabitants passively endure the imposition of particular 'development' schemes and forced incorporation into a neoliberal capitalist logic of production and corresponding forms of ordering the human and the natural realms. Neither are these processes left unaltered in the friction of encounter between the dictates of global capital and local dynamics (Tsing, 2005). As Tsing argues, 'In the historical particularity of global connections, domination and discipline come into their own, but not always in the form laid out by their proponents' (Tsing, 2005: 5).

To further advance understanding of harm vis-à-vis power, it is necessary to turn to the struggles, opposition, and mobilisations that mark the sites of friction in which the politics of harm plays out. Foucault's notion is key here: 'The main objective of these struggles is to attack not so much "such and such" an institution of power, or group, or elite, or class *but rather a technique, a form of power*' (1982: 781, emphasis added).

We can advance criminological understanding of the relation between harm and power via the study of the discourses and on-the-ground encounters in the context of, in this case, palm oil production. The *contested ground* on which the politics of harm plays out demands critical analysis. This kind of approach responds to calls for socially and culturally inclusive critical analysis (for example, White, 2011), not by merely giving a voice to the marginalised, but with the objective of incorporating, studying, and understanding these voices in view of the broader context they are part of, a context of struggle over contested realities and late capitalist mechanisms of ordering the social and natural world.

Notes

The title is taken from a subheading in a brochure disseminated by Fedepalma (2006: 4); translations are mine.

1. Bear in mind that the vegetable oil derived from the fruits of the oil palm serves not only the renewable energy market. In fact, although not addressed here as such, there is a huge market for palm oil's other usages: as cooking oil; as an ingredient in an extraordinary range of food products; in paints, detergents, cosmetics, etc. See Mingorance (2006: 20–22) for an overview of palm oil's multiple uses.
2. I recognise the anthropocentrism that imbues the term 'nonhuman'. For notes on the use speciesist language see, for example, Beirne (2007: 62–64).
3. The concept of 'coloniality' emphasises the continuity of a colonial technology of power, notwithstanding the end of colonial administrations (Castro-Gómez, 2002).
4. The term *feedstock* refers to the raw material that goes into a particular product.
5. As Scott (1998), amongst others, has noted, both the materiality of nature and human organisation intervene with 'ideal' visions of management, manipulation, and control of socio-ecological processes. Global–local dynamics are neither unidirectional, nor imposed on the local without alteration and resistance. Tsing (2005) has powerfully captured this through the concept and analysis of 'friction', of which I will make brief mention in the conclusion to the chapter.
6. I concur with the critique articulated by Cazaux that to speak solely of biodiversity or species loss is an exercise of abstraction that 'passes over the consequences to the well-being of animals as *individual subjects*' (1999: 118). I recognise this as a crucial aspect to be explored and detailed. This is not, however, the focus here.
7. The actual use of and threat of violence, forced displacement, intimidation, and means of land appropriation that readily fit the most conventional definition and perceptions of crime – all intrinsically part of processes of land-use change to agro-industrial models of palm oil production in Colombia (Ocampo Valencia, 2009; Vélez Torres, 2010) – will not be discussed here. Rather, I restrict myself to a brief outline of the adverse social and cultural impacts of palm oil production that illustrate the need for broader analyses of *harm* within criminology.
8. In opposition to the food–fuel complex tied to first-generation agrofuels, the former UN Special Rapporteur on the Right to Food Jean Ziegler, has referred to the severing of food security and sovereignty as 'crimes against humanity' and 'massacre (by) hunger' (Ziegler in Lederer, 2007).

9. This is not to deny cases where agrofuels *have* brought social and economic development and alleviation of poverty. Yet this does not take away from the fact that, overall, palm oil tends to be associated with severe ecological implications and a deepening of levels of material and sociocultural dispossession that people experience.
10. A next thing is how these processes are lived, contested, and transformed in on-the-ground settings; I will open this up to preliminary consideration in the chapter's conclusion as a way to continue the debate.
11. Again, this is (by far) not the only market that drives the large-scale production of palm oil in the global South.
12. Let me underline that I recognise that a persisting coloniality does not exhaust the power relations at play in this context. The colonial is, however, a dimension that in my view demands more explicit attention in green criminological analyses than it has hitherto been granted.
13. By 2020 the planted area should reach 743,000 hectares, with 78 per cent of production geared towards exports. The total area potentially suitable for oil palm cultivation is estimated at 3.5 million hectares (Fedepalma, 2012).
14. Exports dropped significantly in 2010 due to a combination of climatic, plant health, and agronomic factors (Fedepalma, 2011).
15. I do not intend to downplay or altogether reject improvements in the social conditions and living standards brought about by forms of development in this region, but rather to problematise the subjugation and mechanisms of control and domination endured by many of this region's inhabitants under the banner of 'development' and 'social progress', amongst others implemented through such projects as that of the oil palm.
16. To an extent andinocentric imaginary is adhered to by certain Afro-Colombian organizations, too (Arocha Rodríguez and Moreno Tovar, 2007). Moreover it must be noted that oil palm cultivation is not unanimously opposed by all; however, the various arguments adhered to by proponents of palm oil on the one hand and opponents on the other, and the tensions thus produced, are beyond the scope of the discussion here.

References

Alimonda, H. (2011) 'La colonialidad de la naturaleza. Una aproximación a la Ecología Política Latinoamericana', in H. Alimonda (ed.) *La Naturaleza colonizada. Ecología política y minería en América Latina*. Buenos Aires: CLASCO, pp. 21–58.

Altieri, M. (2009) 'Green deserts: monocultures and their impact on biodiversity', in M. S. Emanualli, J. Jonsén and S. Monsalve Suárez (eds) *Red Sugar, Green Deserts. Latin American Report on Monocultures and Violations of the Human Rights to Adequate Food and Housing, to Water, to Land and to Territory*. Halmstad: FIAN International, pp. 67–76.

Arocha Rodríguez, J. and Moreno Tovar, L. del M. (2007) 'Andinocentrismo, salvajismo y afro-reparaciones', in C. Mosquera Rosero-Labbé and C. L. Barcelos (eds) *Afro-reparaciones: Memorias de la Esclavitud y Justicia Reparativa para negros, afrocolombianos y raizales*. Bogotá: Universidad Nacional de Colombia, pp. 587–614.

Beirne, P. (2007) 'Animal rights, animal abuse and green criminology', in P. Beirne and N. South (eds) *Issues in Green Criminology. Confronting Harms against Environments, Humanity and Other Animals*. Cullompton: Willan Publishing, pp. 55–83.

Benton, T. (1998) 'Rights and justice on a shared planet: more rights or new relations', *Theoretical Criminology*, 2(2): 149–175.

Boekhout van Solinge, T. (2010) 'Deforestation crimes and conflict in the Amazon', *Critical Criminology*, 18(4): 263–277.

Cerdas Vega, G. (2009) 'Monocultures and agrofuels: key elements for debate', in M. S. Emanualli, J. Jonsén and S. Monsalve Suárez (eds) *Red Sugar, Green Deserts. Latin American Report on Monocultures and Violations of the Human Rights to Adequate Food and Housing, to Water, to Land and to Territory*. Halmstad: FIAN International, pp. 49–65.

Castro-Gómez, S. (2002) 'The social sciences, epistemic violence, and the problem of the "invention of the Other"', *Nepantla: Views from the South*, 3(2): 269–285.

Cazaux, G. (1999) 'Beauty and the beast: animal abuse from a non-speciesist criminological perspective', *Crime, Law and Social Change*, 31(2): 105–126.

Danielsen, F., Beukema, H., Burgess, N., Parish, F., Brühl, C., Donald, P., Murdiyarso, D., Phalan, B., Reijnders, L., Struebig, M. and Fitzherbert, E. (2008) 'Biofuel plantations on forested lands: double jeopardy for biodiversity and climate', *Conservation Biology*, 23(2): 348–358.

Dauvergne, P. and Neville, K. J. (2009) 'The changing north-south and south-south political economy of biofuels', *Third World Quarterly*, 30(6): 1087–1102.

Escobar, A. (2008) *Territories of Difference: Place, Movements, Life, Redes*. Durham, NC: Duke University Press.

Fedepalma, (2006) *La Agroindustria de la Palma de Aceite en Colombia*. Bogotá: Oficina de Comunicaciones de Fedepalma.

Fedepalma (2011) *Anuario Estadístico/Statistical Yearbook. La agroindustria de la palma de aceite en Colombia y en el mundo/The Oil Palm Agroindustry in Colombia and the World – 2006 – 2010*. Bogotá: Fedepalma.

Fedepalma (2012) *Visión de la Palmicultura Colombiana al 2020*. Available from http://portal.fedepalma.org/vision.htm. Accessed 30 June 2012.

Fitzherbert, E., Struebig, M., Morel, A., Danielsen, F., Brühl, C., Donald, P. and Phalan, B. (2008) How will oil palm expansion affect biodiversity? *Trends in Ecology and Evolution*, 23(10): 538–545.

Flórez López, J. A. and Millán Echeverría, D. C. (2007) *Derecho a la alimentación y al territorio en el Pacífico Colombiano*. Tumaco: Diócesis.

Foucault, M. (1982) 'The subject and power', *Critical Inquiry*, 8(4): 777–795.

Franco, J., Levidow, L., Fig, D., Goldfarb, L., Hönicke, M. and Mendonça, M. L. (2010) 'Assumptions in the European Union biofuels policy: frictions with experiences in Germany, Brazil and Mozambique', *The Journal of Peasant Studies*, 37(4): 661–698.

Halsey, M. (1997) 'Environmental crime: towards an eco-human rights approach', *Current Issues in Criminal Justice*, 8(3): 217–242.

Halsey, M. and White, R. (1998) 'Crime, ecophilosophy and environmental harm', *Theoretical Criminology*, 2(3): 345–371.

Kessler, J. J., Rood, T., Tekelenburg, T. and Bakkenes, M. (2007) 'Biodiversity and socioeconomic impacts of selected agro-commodity production systems', *The Journal of Environment and Development*, 16(2): 131–160.

Lederer, E. M. (2007) Production of biofuels 'is a crime'. *The Independent* [Online], 27 October. Available from: http://www.independent.co.uk/environment/green-living/production-of-biofuels-is-a-crime-398066.html. Accessed 28 February 2012.

Levidow, L. and Paul, H. (2010) 'Global agrofuel crops as contested sustainability, part I: sustaining what development?' *Capitalism Nature Socialism*, 21(2): 64–86.

Lynch, M. (1990) 'The greening of criminology: a perspective on the 1990s', *Critical Criminologist*, 2(3): 1–4, 11–12.

McCarthy, J. and Zen, Z. (2010) Regulating the Oil Palm Boom: Assessing the Effectiveness of Environmental Governance Approaches to Agro-industrial Pollution in Indonesia, *Law and Policy*, 32(1): 153–179.

McMichael, P. (2009) The Agrofuels Project At Large, *Critical Sociology*, 35(6): 825–839.

Mignolo, W. (2000) *Local Histories/Global Designs: Coloniality, Subaltern Knowledges, and Border Thinking*. Princeton: Princeton University Press.

Mingorance, F. (2006) The flow of palm oil Colombia-Belgium/Europe. A study from a human rights perspective. Brussels: Coordination Belge pour la Colombie.

Moore, J. W. (2000) Environmental Crises and the Metabolic Rift in World-Historical Perspective, *Organization and Environment*, 13(2): 123–158.

Mosquera Rosero-Labbé, C. (2007). *Reparaciones* para negros, afrocolombianos y raizales como *rescatados* de la Trata Negrera Transatlántica y *desterrados* de la guerra en Colombia. In: *Afro-reparaciones: Memorias de la Esclavitud y Justicia Reparativa para negros, afrocolombianos y raizales*. Mosquera Rosero-Labbé, C. and Barcelos, C. L. (eds.). Bogotá: Universidad Nacional de Colombia, 213–276.

Ocampo Valencia, S. (2009) Agroindustria y conflicto armado. El caso de la palma de aceite, *Colombia Internacional,* 70: 169–190.

Oslender, U. (2008) Another History of Violence. The Production of 'Geographies of Terror' in Colombia's Pacific Coast Region, *Latin American Perspectives*, 35(5): 77–102.

Pérez-Rincón, M. A. (2008) Los agrocombustibles: ¿sólo canto de sirenas? In: *Agrocombustibles 'Llenando tanques, vaciando territorios'*. Vélez, I. (ed.). Bogotá: CENSAT Agua Viva, pp. 81–116.

Quijano, A. (2007) Coloniality and modernity/rationality, *Cultural Studies*, 21(2–3): 168–178.

Ramírez Vidal, L. A. (2007) Afro-reparaciones en tierras de comunidades negras: Vuelta de Candelilla, Tumaco. In: *Afro- reparaciones: Memorias de la Esclavitud y Justicia Reparativa para negros, afrocolombianos y raizales*. Mosquera Rosero-Labbé, C. and Barcelos, C. L. (eds.). Bogotá: Universidad Nacional de Colombia, 405–421.

Restrepo, E. (2005) Un océano verde para extraer aceite. Hacia una etnografía del cultivo de la palma africana en Tumaco, *Universitas Humanistica*, 58: 72–87.

Scott, J. (1998) *Seeing Like a State. How Certain Schemes to Improve the Human Condition Have Failed*. London: New Yale University Press.

Smolker, R., Tokar, B. and Petermann, A. (2008) *The Real Cost of Agrofuels: Impacts on food, forests, peoples and the climate*. Global Forest Coalition.

South, N. (2007) The 'corporate colonisation of nature': Bio-prospecting, bio-piracy and the development of green criminology. In: *Issues in Green Criminology*.

Confronting harms against environments, humanity and other animals. Beirne, P. and South, N. (eds.). Cullompton: Willan Publishing, 230–247.

Tsing, A. L. (2005) *Friction: An Ethnography of Global Connection*. Princeton: Princeton University Press.

Tuhiwai Smith, L. (1999) *Decolonizing Methodologies. Research and Indigenous Peoples*. London: Zed Books Ltd.

Vélez Torres, I. (ed.) (2010) *Misión internacional para la verificación del impacto de los agrocombustibles en 5 zonas afectadas por los monocultivos de palma aceitera y caña de azúcar en Colombia*. Bogotá: CENSAT Agua Viva.

Vélez, H. and Vélez, I. (2008) Los espejismos de los agrocombustibles. In: *Agrocombustibles 'Llenando tanques, vaciando territorios'*. Edited by I. Vélez. Bogotá: CENSAT Agua Viva, 15–58.

Ventes, J., Roa Avendaño, T. and Toloza, J. (2008) Dinámicas de una resistencia cantada. In: *Agrocombustibles 'Llenando tanques, vaciando territorios'*. Edited by I. Vélez. Bogotá: CENSAT Agua Viva, 181–192.

Walsh, C. (2007) Shifting the geopolitics of critical knowledge. Decolonial thought and cultural studies 'others' in the Andes, *Cultural Studies*, 21(2–3): 224–239.

Walters, R. (2010) Toxic Atmospheres Air Pollution, Trade and the Politics of Regulation, *Critical Criminology*, 18(4): 307–323.

White, R. (2002) Environmental Harm and the Political Economy of Consumption, *Social Justice*, 29(1–2): 82–102.

White, R. (2007) Green criminology and the pursuit of social and ecological justice. In: *Issues in Green Criminology. Confronting harms against environments, humanity and other animals*. Beirne, P. and South, N. (eds.). Cullompton: Willan Publishing, 32–54.

White, R. (2011) *Transnational Environmental Crime. Towards an eco-global criminology*. London: Routledge.

Wilcove, D. S. and Koh, L. P. (2010) Addressing the threats to biodiversity from oil-palm agriculture, *Biodiversity and Conservation*, 19(4): 999–1007.

Ziegler, J. (2007) *Report of the Special Rapporteur on the right to food*. United Nations General Assembly, sixty-second session, A/62/289.

Index

Made in the USA
Monee, IL
13 September 2026

62976736R00154